Municipal Solid Waste Management in Developing Countries

Municipal Solid Waste Management in Developing Countries

Sunil Kumar

CRC Press
Taylor & Francis Group
Boca Raton London New York

CRC Press is an imprint of the
Taylor & Francis Group, an **informa** business

CRC Press
Taylor & Francis Group
6000 Broken Sound Parkway NW, Suite 300
Boca Raton, FL 33487-2742

First issued in paperback 2020

© by Taylor & Francis Group, LLC
CRC Press is an imprint of Taylor & Francis Group, an Informa business

No claim to original U.S. Government works

ISBN-13: 978-0-367-57428-4 (pbk)
ISBN-13: 978-1-4987-3774-6 (hbk)

**Visit the Taylor & Francis Web site at
http://www.taylorandfrancis.com**

**and the CRC Press Web site at
http://www.crcpress.com**

Dedicated to My Father

Contents

Foreword xiii
Preface.. xv
Acknowledgments xvii
Author.. xix
Contributors... xxi

1. Overview 1
   1.1   Introduction 1
   1.2   Municipal Solid Waste Management: Urban Problem 2
   1.3   Various Facets of MSWM.. 3
   1.4   Need for Integrated Management of Municipal Solid Waste......... 4
         1.4.1   Why Focus on IMSWM Systems?.................................... 6
         1.4.2   Research Proving Need for Integrated Management
                 of MSW 6
         1.4.3   Comparison between Developing and Developed
                 Countries in the Integrated Management of MSW Sector.... 8
   1.5   Structure of the Book.. 9

2. MSW and Its Management 11
   2.1   Definition of Waste, Solid Waste, and MSW 11
         2.1.1   Mechanism of MSW Generation 11
         2.1.2   MSW Composition and Characteristics 12
                 2.1.2.1   Waste Characteristics.................................... 14
                 2.1.2.2   Physical Characteristics................................. 14
                 2.1.2.3   Chemical Characteristics................................ 15
                 2.1.2.4   Proximate Analysis 16
                 2.1.2.5   Ultimate Analysis.. 16
   2.2   MSWM: Functional System 17
         2.2.1   Historical Development of MSWM System 17
         2.2.2   Evolution of MSWM System.. 18
         2.2.3   Modern MSWM System Techniques in Developing
                 Countries... 18
                 2.2.3.1   Techniques Available for Implementation of
                           MSWM Systems in Developing Countries......... 20
                 2.2.3.2   Status of MSWM Systems in Developing
                             Countries together with Recent Available
                             Technologies... 20
                 2.2.3.3   Basic Comparison between Available
                             Technologies Used in Developing and
                             Developed Countries 23

            2.2.3.4   Problems to Be Faced while Implementing
                      MSWM Systems in Developing Countries.........25
            2.2.3.5   Categories of Problems Common to Waste
                      Management in Developing Countries..............27

3. Generation of MSW 29
   3.1   Source of Generation of MSW 29
   3.2   Factors Affecting Rate of Generation of MSW............................32
   3.3   MSW Generation Scenarios in Developing Countries 36

4. Management Practices of MSW in Developing Countries 39
   4.1   Introduction 39
   4.2   Legislation and Laws in MSW in India 40
         4.2.1   New MSW Draft Rule 2015...40
   4.3   Role of the Municipal Corporation...49
   4.4   Role of Ragpickers in MSWM 50

5. Component Technologies for Municipal Solid Waste
   Management 53
   Mukesh Kumar Awasthi, Amanullah Mahar, Amjad Ali, Quan Wang,
   and Zengqiang Zhang
   5.1   Introduction 53
   5.2   Collection of Municipal Solid Waste 54
         5.2.1   House-to-House Collection 54
         5.2.2   Community Bin System 55
   5.3   Transportation 55
         5.3.1   Technical Requirements of Municipal Solid Waste
                 Transport Vehicles 56
         5.3.2   Types of Municipal Solid Waste Transportation
                 Vehicles 57
         5.3.3   Transfer Stations 57
         5.3.4   Optimization of Transportation Routes 57
   5.4   Biological and Thermal Processing Methods 58
         5.4.1   Composting 58
                 5.4.1.1   Chemical Transformations during
                           Composting.. 60
                 5.4.1.2   Design of Compost Mixtures 60
                 5.4.1.3   Components of Compost Mix............................ 61
                 5.4.1.4   Types of Feedstock 61
                 5.4.1.5   Monitoring and Parameter Adjustment..............63
                 5.4.1.6   Main Types of Composting System.....................67
         5.4.2   Biomethanation 71
                 5.4.2.1   Different Biochemical Processes of
                           Biomethanation..73
                 5.4.2.2   Parameters Affecting Anaerobic Digestion........74

| | | 5.4.2.3 | Types of Anaerobic Digester | 75 |

5.4.2.3 Types of Anaerobic Digester 75
    5.4.3 Thermal Processing of Municipal Solid Waste 75
        5.4.3.1 Incineration 76
        5.4.3.2 Refuse-Derived Fuel 79
        5.4.3.3 Pyrolysis 80
5.5 Reuse and Recycling 81
5.6 Ultimate Disposal Methods 82
    5.6.1 Types of Landfill 82
    5.6.2 Environmental Impacts of Landfilling and Their
        Control 83
    5.6.3 Sanitary Landfilling with Biogas Recovery 84
        5.6.3.1 Methods of Sanitary Landfilling 84
        5.6.3.2 Leachate Collection System 85
        5.6.3.3 Biogas Recovery from Landfill 87
    5.6.4 Carbon Storage in Landfill 88

6. Kinetics of Waste Degradation 89
Poornima Jayasinghe and Patrick Hettiaratchi
6.1 Waste Degradation Process 89
6.2 Aerobic Waste Degradation Process 89
    6.2.1 Initial Mesophilic Phase 91
    6.2.2 Thermophilic Phase 91
    6.2.3 Secondary Mesophilic Phase 91
    6.2.4 Stoichiometric Equation for Aerobic Waste
        Degradation 91
    6.2.5 Factors Affecting Aerobic Degradation 92
        6.2.5.1 Moisture 92
        6.2.5.2 Feedstock Composition (C:N ratio) 92
        6.2.5.3 Oxygen Requirements 92
        6.2.5.4 pH 93
6.3 Anaerobic Waste Degradation Process 93
    6.3.1 Hydrolysis Stage 93
    6.3.2 Acetogenesis Stage 94
    6.3.3 Methanogenesis Stage 95
    6.3.4 Stoichiometric Equation for Anaerobic Waste
        Degradation and Estimation of Theoretical Methane
        Yield 96
    6.3.5 Factors Affecting Anaerobic Waste Degradation 97
        6.3.5.1 pH 97
        6.3.5.2 Moisture Content 98
        6.3.5.3 Temperature 98
        6.3.5.4 Nutrients 98
        6.3.5.5 Inhibitors 98
6.4 Waste Degradation Sequence in Landfills 99
6.5 Waste Degradation in Anaerobic Digesters 100

6.5.1   Moisture Controlled 101
6.5.2   Temperature Controlled.. 101
6.5.3   Feedstock/Substrate Delivery Controlled....................... 101
6.6   Waste Degradation Reaction Kinetics.. 102
6.6.1   Anaerobic Waste Degradation: Kinetic Model 102
6.6.1.1   Solid Hydrolysis 103
6.6.1.2   Intermediate Waste Degradation: Aqueous
Carbon.. 106
6.6.1.3   Formation of Methane 108
6.7   Aerobic Waste Degradation: Kinetic Model................................ 109
6.8   Landfill Gas Generation Kinetics 110
6.8.1   Scholl Canyon Model and USEPA LandGEM................. 110
6.8.2   IPCC First-Order Decay Model... 113
6.8.3   IPCC Default Method 113
6.8.4   Triangular Model.. 114
6.9   Anaerobic Digester Kinetics.. 115
6.9.1   Batch Digester.. 116
6.9.2   Continuous Flow Digester... 116
6.9.3   Semibatch Digester 117

7. Systems Approaches in Municipal Solid Waste Management 119
7.1   Introduction 119
7.2   Development Drivers for Solid Waste Management 120
7.2.1   Solid Waste Management Development in High-
Income Countries... 120
7.2.2   Municipal Solid Waste Management in Low- and
Medium-Income Countries 122
7.2.2.1   Urbanization 123
7.2.2.2   Cultural and Socioeconomic Aspects.............. 123
7.2.2.3   Political Landscape 124
7.2.2.4   International Influences 125
7.3   Need for Systems Approaches to Solid Waste Management...... 125
7.4   Integrated Solid Waste Management: Systems Perspective........ 126
7.5   Systems Approaches.. 126
7.6   Systems Engineering Principles... 127
7.6.1   Systems Definition.. 127
7.6.2   Systems Thinking 128
7.6.3   Systems Engineering Approaches..................................... 128
7.7   System-of-Systems Approach... 129
7.8   Centralized and Decentralized Systems 130
7.9   Systems Analysis Techniques for Municipal Solid Waste
Management 131
7.9.1   Systems Analysis Techniques 131

8. Municipal Solid Waste Management Planning 133
  8.1   Introduction 133
  8.2   Effects of Improper Planning for Implementation of MSWM
        Systems 133
  8.3   Requirements in MSW Planning 134
        8.3.1   Factors Affecting IMSWM 135
        8.3.2   Planning IMSWM Systems 135
  8.4   Tactical and Strategic Planning for Implementation of
        MSWM Systems 137
        8.4.1   Strategic Planning of MSWM Systems 137
  8.5   Long- and Short-Term Planning for MSWM Systems 140
  8.6   Basic Planning Model 144

9. Models for Municipal Solid Waste Management Systems 145
  9.1   Models for Community Bin Collection Systems 145
  9.2   Models for Vehicle Routing 147
  9.3   Landfill Gas Modeling 148

References 153
Index 169

Foreword

सी.एस.आई.आर. - राष्ट्रीय पर्यावरण अभियांत्रिकी अनुसंधान संस्थान
CSIR - National Environmental Engineering Research Institute
(वैज्ञानिक तथा औद्योगिक अनुसंधान परिषद् / Council of Scientific & Industrial Research)
(वैज्ञानिक तथा औद्योगिक अनुसंधान विभाग, विज्ञान एवं प्रौद्योगिकी मंत्रालय, भारत सरकार के अंतर्गत स्वायत्त संगठन)
(Autonomous Organisation under the Dept. of Scientific and Industrial Research, Ministry of Science & Technology, Govt. of India)

NEERI

Solid Waste Management (SWM) has become one of the crucial parameters of urbanization. India bas recognized its importance and has set-up "Clean India Mission". Time has come to move forward and to emphasize on all the different aspects of SWM through proper assessment and providing appropriate solution. Scientific understanding is key to achieve success in this field.

This book entitled *Municipal Solid Waste Management in Developing Countries* covers all the aspects of SWM of developing countries. It provides the basic guidelines to achieve sustainable Municipal Solid Waste Management (MSWM) in developing countries.

The relevance of knowledge in managing municipal solid waste has been brought out through many subtopics. The in-depth discussion and peer reviewed write up on case studies have made the book the right material for researchers, academicians, students as well as decision makers.

I am sure this book is useful to all stakeholders working in the field of solid waste management in developing countries and that it will provide insight to others.

Dr. Rakesh Kumar
Director, NEERI

Preface

Rapid urbanization in major cities of the developing as well as the developed world puts immense pressure on the existing infrastructural facilities, among which one such sector is municipal solid waste management (MSWM). Existing infrastructures are not adequate to deal with the increasing quantity of municipal solid waste (MSW). The major reason behind the poor project planning and implementation of rules and regulations in developing nations is the lack of knowledge about the various available technologies, the best management practices, and the fundamentals of MSWM. To achieve a sustainable MSWM system in a developing nation, it is necessary to develop appropriate waste collection, treatment, recycling, and disposal methods, and knowledge of various treatment technologies. Developing a proper system requires the involvement of all aspects of planning and designing an integrated solid waste management (ISWM) system, such as technical, political, environmental, socioeconomic, legal, and cultural aspects.

This book is intended to cover all the fundamental concepts of MSWM, the various component systems, such as collection, transportation, processing, and disposal, and their integration. This book also covers the various component technologies available for the treatment, processing, and disposal of MSW. A successful MSWM system is the result of proper planning and strategy. This book focuses on various systems approaches and planning techniques for MSWM in low- and medium-income countries. The comprehensive details are covered in nine chapters. The first three chapters provide a basic overview of MSWM composed of the mechanism of MSW generation, and its composition, characteristics, and scenarios in developing countries. Chapter 4 covers MSWM practices in developing countries, laws, legislations, and the role of municipal corporations and ragpickers in MSWM. Chapter 5 discusses collection and transportation methods for MSW and provides detailed information on various biological and thermal processing methods, such as composting, biomethanation, incineration, pyrolysis, and so on. Chapter 5 also covers reuse and recycling techniques and disposal methods such as sanitary landfilling, biogas recovery from landfill, and the environmental impacts of landfill. Chapter 6 provides a thorough explanation of aerobic and anaerobic waste degradation, its reaction kinetics, and its stoichiometry. It also provides information on waste degradation sequences and gas generation in a landfill. Chapter 7 provides details about the various development drivers for MSWM and challenges in low- and medium-income countries; then it focuses on the need for systems approaches and ISWM perspectives. In addition to this, Chapter 7 also elaborates on various systems engineering principles and approaches along with a comparison of centralized and decentralized waste management systems. In Chapter 8, some important aspects of planning are discussed in detail. Chapter 9 discusses various models for community bin systems, vehicle routing, and landfill gas modeling.

This book is written in view of actual scenarios in developing countries and provides knowledge to develop solutions for prolonged problems in these nations. It is written mainly for undergraduate and postgraduate students, research scholars, professionals, and policy makers. This book ultimately provides the basic guidelines for a holistic approach to achieving sustainable MSWM systems in developing countries.

Dr. Sunil Kumar
Senior Scientist
Solid and Hazardous Waste Management Division
CSIR-NEERI
Nehru Mrag
Nagpur, India
www.neeri.res.in
sunil_neeri@yahoo.co.in

Acknowledgments

This book, entitled *Municipal Solid Waste Management in Developing Countries*, consists of comprehensive information on all the components of municipal solid waste management in developing countries. I wish to express my sincere gratitude for the active support extended by all my fellow colleagues in the Solid and Hazardous Waste Management Division of the Council of Scientific and Industrial Research-National Environmental Engineering Research Institute (CSIR-NEERI), Nagpur, India, in completing this book. I also wish to express thanks to my wife, Rashmi Rani, and both of my sons, Master Vimal Kumar and Master Sourabh Kumar, for sparing time to help me complete the book in time.

I am thankful to Dr. Rakesh Kumar, Director, CSIR-NEERI, for giving permission to publish this book. I express my special thanks to all my students—namely, Rena, Shashi Arya, Anurag Gupta, Abhishek Khapre, Hiya Dhar, and Avick Sil—for helping me in finalizing this book. I am thankful to all those who have directly or indirectly contributed to the publication of this book.

Dr. Sunil Kumar
Senior Scientist
Solid and Hazardous Waste Management Division
CSIR-NEERI
Nehru Mrag
Nagpur, India
www.neeri.res.in
sunil_neeri@yahoo.co.in

Author

Sunil Kumar is a senior scientist in the Solid and Hazardous Waste Management Division at the Council of Scientific and Industrial Research's National Environmental Engineering Research Institute (CSIR-NEERI), Nagpur, India. For the last 15 years, he has been working in the field of solid and hazardous waste management and has authored numerous scientific papers and edited several special issues for reputed national and international journals.

Contributors

Amjad Ali
College of Natural Resources and
   Environment
Northwest A&F University
Yangling, PR China

Mukesh Kumar Awasthi
College of Natural Resources and
   Environment
Northwest A&F University
Yangling, PR China
and
Department of Biotechnology
Amicable Knowledge Solution
   University
Satna, India

Patrick Hettiaratchi
Department of Civil Engineering
Center for Environmental
   Engineering Research and
   Education
University of Calgary
Calgary, Alberta, Canada

Poornima Jayasinghe
Department of Civil Engineering
Center for Environmental
   Engineering Research and
   Education
University of Calgary
Calgary, Alberta, Canada

Amanullah Mahar
Centre for Environmental Sciences
University of Sindh
Jamshoro, Pakistan

Quan Wang
College of Natural Resources and
   Environment
Northwest A&F University
Yangling, PR China

Zengqiang Zhang
College of Natural Resources and
   Environment
Northwest A&F University
Yangling, PR China

1

Overview

1.1 Introduction

Globally, solid waste is one of the subjects of greatest concern for the environment. Historically, the environment has been considered a sink for all the waste produced by human activities since the beginning of civilization. The suitable environment in which the human race resides, along with the flora and fauna, is a gift of nature, formed through different phenomenal natural activities. Today's technologically advanced society, in the wake of urbanization and the advancement of civilization, has knowingly or unknowingly contributed substantially to altering the natural environment. To satisfy daily needs, human activities contribute to all types of pollution related to air, water, and land. As air and water comprise an essential part of the human being, these effects are easily noticed. Therefore, numerous attempts are being made to prevent and control pollution.

One of the issues of greatest concern is the impact of environmental pollution caused by solid waste. However, this is still not given much focus despite the fact that much more will have to be done in developed countries to combat the situation. It is also problematic in developing countries.

Solid waste management is a common term for *garbage management*. It consists of discarded leftovers of the materials used in daily life; it can be food waste, metals, and so on. Solid wastes are classified into many categories based on their source, origin, and type of waste. Municipal solid waste (MSW) is quite different from other types of waste as it does not have a definite time frame of generation. Each and every moment, waste is generated at different locations and, to handle the chaos and nuisance of MSW, an organized process for storage, collection, segregation, transportation, processing, and the disposal of waste is being implemented through a managed channel called the *MSW management system*.

MSW is produced and handled everywhere in the world and different countries are moving at different speeds to set and meet management targets. MSW management (MSWM) in developing countries poses critical challenges as it grapples with sudden and rapid urbanization. The globalization of the economy has resulted in the availability of a greater variety of products, causing the generation of an enormous amount of solid waste.

1

There is no comprehensive data on the rate of waste generation, collection, coverage, storage, transport, or disposal volumes. For the massive generation of waste, management practice in developing countries is not up to the mark. Attempts at the betterment of these systems, mostly by borrowing Western technologies, has met with little success, mainly because of the inappropriateness of such adopted technologies in the present conditions. Thus, developing countries need a basic guideline through which they can manage their MSW.

1.2 Municipal Solid Waste Management: Urban Problem

Urban areas are called a *place of hope*. They form hubs offering several facilities, such as medical access, cafeterias, restaurants, jobs, and industries to manufacture different products. People from rural areas, in search of a better lifestyle and money, flock to urban places to find a better standard of living. The floating population, along with the urban population, leads to enormous waste generation. Factors, such as population in any given area and consumption patterns determine the generation of waste. An urban population is divided into four categories:

- High-income group
- Middle-income group
- Low-income group
- Slums

Each of these groups has a different way of living and intake of food. The generation of waste not only depends on the food habits and lifestyle of people, but also on the abundance and type of a region's natural resources. These factors ultimately lead to enormous waste generation, thus making the MSWM a must for urban bodies.

MSWM in urban regions in developing countries faces serious managerial issues as it has specific characteristics related to limited land area and low financial capacity. According to the Asian Development Bank Report (2014) on solid waste management in the Pacific, it is estimated that the region's per capita MSW generation rate is 0.45 kg/capita/day followed by paper, plastic, metal, and glass collectively contributing a further 30% or more by weight.

The Human Development Report (UNDP, 2010) indicates that accessibility to water and sanitation in developing and less developed countries is not encouraging at all. It is found that, in 45% of countries, the population lacks proper sanitation and infrastructure. In addition, the report highlights that the percentage of the population living on degraded land is increasing to an average

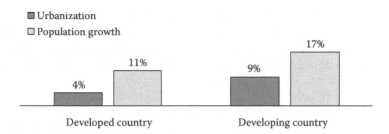

FIGURE 1.1
Population and urbanization growth 1990–2010. (From Khatib, A.I., Municipal solid waste management in developing countries: Future challenges and possible opportunities. In S. Kumar (Ed.), Volume II, *Integrated Waste Management*, InTech, 2011.)

that exceeds 15%. Hence, the consequences of unplanned urbanization growth will certainly lead to significant problems for governments especially with regard to meeting the demand for proper and healthy municipal services. The growth will result in an increase in the quality and complexity of generated waste and overburdens to waste management systems, including solid wastes, and MSW, in particular. Figure 1.1 shows the population and urbanization growth in developed and developing countries for the years 1990–2010.

1.3 Various Facets of MSWM

MSW is a multidisciplinary activity that is not only based on urban and regional planning, but also involves engineering principles along with economic and social issues. The management of MSW starts from the point of generation up to disposal. To carry out the vast task of MSWM, the system requires resources in the form of manpower and machinery for MSW handling operations, such as collection, transportation, loading, and unloading of waste. Vehicles and heavy earthmoving equipment are needed at landfill sites for its ultimate disposal. Another concern for the management of MSW is that it should be designed, planned, and operated in such a way that there should be minimal damage to the environment.

For effective and efficient MSWM, judicious use of heterogeneous technologies is needed so that environmental pollution can be minimized. A dedicated team of experts coordinates, plans, and executes the system, taking all aspects into account: mechanical engineering can be used for better handling and processing equipment; chemical engineering can be used to find better processes and technologies for effective treatment methods; and social science can be used for creating awareness and better public cooperation. There is also a need for economists to build sustainable economic models and for environment engineers to manage the system. Above all, requirements are met by the use of available resources and technologies. The management of MSW depends on the quantity and quality of the waste and the characteristics of MSW vary from place to place. This variation is due to food habits, culture, climate, and so on.

The quantity of solid waste depends on the gross national product (GNP)/capita. The quality is assessed both physically and chemically. Physically, it covers paper, plastics, glass, and organics; chemically, it specifies the percentage composition of protein, carbohydrates, and lignin along with the elemental composition—that is, the percentage of carbon, nitrogen, and phosphorous. As the population is increasing day by day and the generation of MSW is directly proportional to the population, this field requires proper management. However, in this endeavor, MSW is regarded as a low priority. The effect of MSW is not direct but results from its improper management causing a public health hazard and affecting aesthetic values.

A public health hazard due to MSW arises from the breeding of disease vectors, primarily flies and rats. Improper management leads to infestations of insects and rodents. The most conspicuous environmental damage is aesthetic, caused by street littering and unmanaged dumping of waste. Though solid waste has been given less priority, it is coincidentally connected with all other types of pollution. During the processing of waste, such as by incineration, noxious fumes emerge, causing air pollution. Unmanaged and dumped waste at landfills causes leachate, which washes out during the rainy season and floats down to nearby water bodies, ultimately polluting the drinking water. Thus, MSWM is a problem that also illustrates a deep relationship with other kinds of pollution.

MSWM has yet to gain much recognition in developing countries. Because they have limited resources and limited access to technology, developing countries cannot provide adequate techniques to overcome the management issue. In most developing countries, the hazards of pollution have been realized and people have started thinking in terms of better technologies and low-cost efficient systems that would suit their profile. Continuous interaction of society with the system has taken shape in various ways and is available in different forms in various parts of the world. This evolutionary study will provide us with ideas to formulate particular systems in context and in conjunction with the available resources.

1.4 Need for Integrated Management of Municipal Solid Waste

The increasing volumes of waste that are being generated would not be a problem if the waste was viewed as a resource and managed properly (UNEP 2001).

Integrated solid waste management (ISWM) is a complete strategy of prevention, recycling, and the proper management of solid waste. This comprehensive program is ultimately helpful for protecting human health and the environment. ISWM assesses local need and conditions to map out the most appropriate waste management practice for particular conditions. The major activities are complete waste management including composting, combustion, and disposal in a properly designed and managed landfill site (USEPA 2005).

The tremendous increase in the amount and the different categories of waste—as a result of continuous economic growth, urbanization, and industrialization—is becoming a growing problem for national and local governments who are aiming to ensure its effective and sustainable management. It is estimated that, in 2006, the total amount of MSW generated globally reached 2.02 billion tons, representing a 7% annual increase since 2003 (Global Waste Management Market Report 2007). It is also observed that between 2007 and 2011, the global generation of MSW has risen by 37.3%—equivalent, roughly, to an 8% increase per year (Muzenda et al. 2012).

Thus, the amount of MSW generated in the world is steadily increasing and every government is currently concentrating on the best way to significantly reduce the management issues related to solid waste management (Khan 2014). There is a need to implement integrated management of MSW globally.

Now when it comes to the implementation of MSWM strategies in developed and developing countries, serious statistics of their implications are carried out. In developed countries, several methods and plans for executing MSWM are in operation. However, some sectors are having difficulties in carrying out their integrated MSWM programs. There are still major gaps to be filled in this area.

The World Bank estimates that developing countries are spending 20%–80% of their available funds on solid waste management (SWM). However, open dumping and open burning is quite a normal phenomenon in such countries: 30%–60% of urban solid waste is not collected, thus waste collection and waste management systems benefit only 50% of the population. Of the total budget for MSW management 80%–90% is spent on MSW collection in low income countries. High income countries spent less than 10% of their total budget. (Unnisa et al. 2013). Less than 10% of the budget is collected in high-income countries for SWM, enabling significant funds to be used for different treatment facilities (Kadafa et al. 2012). Concerted public participation in advanced countries helps to reduce the collection cost, thereby facilitating recycling and recovery (United Nations Economic and Social Commission for Asia and the Pacific 2009).

Developing countries face a critical challenge to the management of their waste. The aspect of greatest focus is to reduce the final volume of waste to generate sufficient funds. The final volume of waste could be substantially reduced if the waste is channeled for material and resource recovery. Thus, recovered material and resource can be used to generate financial support for waste management (UNEP 2009). This forms the premise for an integrated municipal solid waste management (IMSWM) system based on the 4R (reduce, recycle, recover, and reuse) principle.

It is necessary to know and understand the reasons for the need to implement the IMSWM program in waste management practices. These are described in the following sections. Developing and implementing IMSWM is dependent on having the following information:

- Comprehensive data on present and anticipated waste situations
- Supportive policy frameworks
- Knowledge and capacity to develop plans/systems
- Proper use of environmentally sound technologies
- Appropriate financial instruments to support its implementation

1.4.1 Why Focus on IMSWM Systems?

The expansion of a city is identified by increases in its financial and developmental activities and these entirely depend on lifestyle, abundance, type of natural resources, and consumption pattern. As consumption patterns and lifestyles have changed significantly, the characteristics of waste have changed (Toolkit for ISWM Planning 2007).

An inability to fully grasp the problem of waste generation has resulted in MSW being one of the most challenging problems of urban environmental degradation.

There is also an absolute necessity to integrate the informal sector (which consists of rag pickers and illegal or unauthorized recyclers) into the mainstream waste management process as they handle a substantial amount of waste without the mandatory environmental safeguards.

A probable solution to waste management would be an integrated approach that would incorporate collective management of all types of waste and the implementation of the 4R policies and strategies (Developing Integrated Solid Waste Management Plan 2009). Figure 1.2 shows the need for an IMSWM system.

1.4.2 Research Proving Need for Integrated Management of MSW

The world population is increasing daily and projections for 2015 are nearly 7.2 billion (UNEP 2005). It is estimated that, at the current rate, rapid urbanization would result in two-thirds of the world's people living in cities by 2025. In fact, every day, the urban population in developing countries grows by more than 150,000 people (United Nations Department of Economic and Social Affairs [UNEDSA] 2005). Although rapid urbanization itself is not

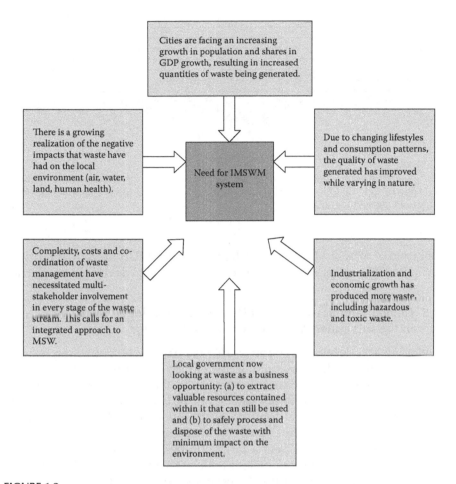

FIGURE 1.2
The need for ISWM. (From United Nations Environment Program (UNEP). *Assessment of Current Waste Management System and Gaps Therein*, Volume 2, 2009.)

necessarily a problem, haphazard and spontaneous growth could result in many environmental problems, such as encroachment of public space and riverbank, air pollution, and water pollution, along with the generation of tremendous amounts of solid waste (UNEP 2001). MSW is the most complex heterogeneous solid waste stream, in contrast with more homogeneous waste streams, such as industrial or agricultural waste (Wang and Nie 2001). In cities, even a slight increase in income can cause people's consumption patterns to change (Medina 1997), which results in waste types and quantities that pose a greater challenge for the municipalities to handle. For example, a study in India showed that a 49% increase in population results in the

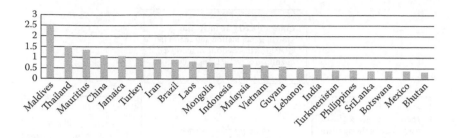

FIGURE 1.3
MSW generation rates (kg/capita/day) for 23 Organisation for Economic Co-operation and Development (OECD) developing countries. (From Troschinetz and Mihelcic, *Waste Management*, 29(2), 915–923, 2009.)

generation of 67% of MSW during the same period (UNEP 2001). Figure 1.3 shows the MSW generation rate (kg/capita/day) for 23 developing countries.

1.4.3 Comparison between Developing and Developed Countries in the Integrated Management of MSW Sector

Developing and developed countries display particular contrasting features in the name of recycling research. In developed countries, the characterization of MSW has been conducted, along with the generation of waste and recovery rates, by accurate databases such as the United Nations Environment Programme (UNEP), Global Environment Outlook (GEO), DataPortal (DP), and World Research Institute Earth (WRIE) trends. It has been found that developed countries implement a large number of industrialized recycling activities and these are more or less removed from the daily life of a citizen by their sophisticated curbside recycling programs. They also focus on the exact tools, models, and policy analysis, such as command-and-control and social-psychological and economic incentives (Taylor 2000). Developing countries focus less on understanding the recycling research as practiced in developed countries and more on practical and direct factors that influence the institutions and on basics associated with MSWM (Troschinetz and Mihelcic 2009).

Developed countries typically use curbside recycling programs to collect and segregate waste for recycling. On the other hand, developing countries employ the social sector known as *rag pickers* to handle waste. Rag pickers are people who pick up rags and other recyclable wastes. This is labor-intensive, low-technology, low-paid, and unregulated work. The waste collected by the rag pickers is then sold to recycling shops, middlemen, or exporters. Sometimes, people view scavengers as a nuisance (Sverige 2014).

1.5 Structure of the Book

The chapters of the book comprehensively cover the entire management issue of MSW in developing countries. Each chapter starts with fundamentals along with additional data to offer complete cohesiveness.

- Chapter 1 gives a general overview and other information, such as problems in existing MSWM in urban areas. Special emphasis has been placed on the integrated municipal solid waste management system (IMSWMS).
- Chapter 2 deals with MSWM, starting with general information about waste and MSW. This chapter also includes functional systems comprising historical development, the evolution of MSW along with the modern MSWM system, and the problems of management.
- Chapter 3 focuses on explaining the generation of MSW along with the role of a population in MSW generation. This chapter also focuses on a scenario regarding the generation of MSW in a developing country and also includes the rate of generation of solid waste in the last 10 years in the developing country.
- Chapter 4 focuses on the management practices of MSW, legislation and laws on MSW along with the role of a rag picker in the MSWM system in developing countries.
- Chapter 5 deals with technical aspects of MSWM, focusing on collection, transportation, and processing in a detailed manner.
- Chapter 6 deals with the chemistry concerned with the mass balance equation for aerobic degradation, anaerobic degradation, the stoichiometry of aerobic and anaerobic degradation along with the computation of methane, ammonia, and so on. This chapter examines how rapid chemical reaction occurs in MSW.
- Chapters 7 and 8 cover the system approach for MSWM and planning. These two chapters emphasize the practicalities of the two major issues that are very much required to understand the system.
- Chapter 9 focuses on models for MSWM along with case studies. Detailed information about models for community bins, vehicle routing, and landfill gas modeling is given.

2

MSW and Its Management

2.1 Definition of Waste, Solid Waste, and MSW

Any material is termed *nonusable* after its complete utilization for an individual purpose is called *waste*. Waste is generated either by human beings, animals, or plants, or from any natural or artificial process. Waste takes many forms, such as municipal solid waste (MSW), biodegradable waste, nonbiodegradable waste, chemical waste, construction and demolition waste, electronic waste, biomedical waste, wastewater, sludge, toxic waste, industrial waste, food waste, and so on.

Any garbage or refuse found in a solid or semisolid physical state is termed *solid waste*. Sources of solid waste include households, public places, commercial institutions, hospitals, industries, semisolid waste from wastewater plants, electronic industries, and so on. Various sources of solid waste generation are presented in Table 2.1.

MSW is trash that consists of everyday items useful for public use that have been discarded, such as plastics, paper, rags, green waste, electronic waste, inert waste such as construction and demolition debris, and so on.

MSW generally consists of three types:

- Residential waste that is generated by individual households located in inland areas
- Commercial waste generated from large single sources such as schools, colleges, and hotels
- Waste from municipal services, such as streets, public gardens, and so on

2.1.1 Mechanism of MSW Generation

Waste produced from various sources such as households, commercial buildings, street sweeping, and parks is first collected and stored at a primary storage point, such as street containers. Then the waste is transferred to secondary storage points in cities where waste is stored and segregated

TABLE 2.1

Sources of Solid Waste Generation in a Community

Sources	Location of Generation of Waste	Types of Solid Waste
Residential	Low-, medium-, and high-rise apartments, houses	Food waste, paper, textile, plastics, glass–aluminum, tin cans, street leaves etc.
Commercial institutions	Restaurants, offices, hotels, markets, schools, hospitals	Cardboard, plastics, food waste, metal waste, consumer electronics
Industrial (no process waste)	Construction, fabrication, light and heavy manufacturing, refineries, power plants	Paper, cardboard, plastics, wood, glass, ashes, hazardous waste
Construction and demolition	Road repair sites, new construction sites	Wood, steel, concrete, stone
Municipal services (excluding treatment facilities)	Parks, beaches, recreational areas, street cleaning	General waste from parks, street sweepings
Treatment plant sites	Water, waste water, and industrial treatment process	Treatment plant waste, principally composed of residual sludge and other residual material
Agricultural waste	Field and row crops, vineyards, dairies, farms	Spoiled food waste, agricultural waste

Source: Tchobanoglous et al., *Integrated Solid Waste Management Engineering Principles and Management Issues.* New York: McGraw-Hill, 1993.

before it is sent for processing and final disposal. Waste is divided into three categories:

- Inert (sand, stones, etc.)
- Organics (hydrocarbons) separated into two groups, that is, high-moisture biodegradable (kitchen waste) and low-moisture organics (polythene, rubber tires, etc.)
- Recyclables

After that, the inert portion of the waste is sent to landfill for final disposal, and organics are sent to various processing options, such as composting, biomethanation, incineration, pyrolysis, and so on. The typical mechanism of MSW generation is shown in Figure 2.1.

2.1.2 MSW Composition and Characteristics

To prepare a plan for a municipal solid waste management (MSWM) system for any local government, it is necessary to carry out an assessment of the waste stream. Waste stream assessment gives necessary information about

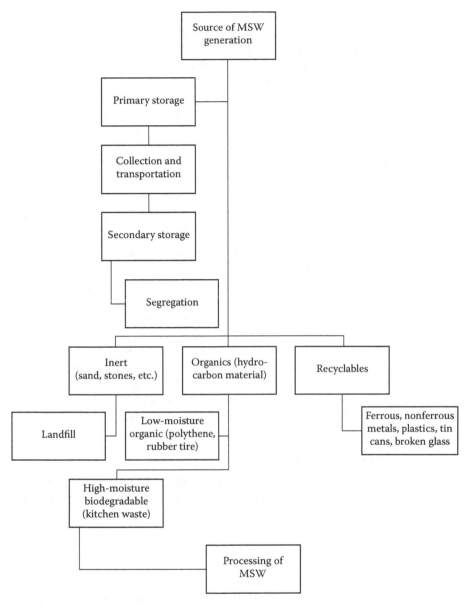

FIGURE 2.1
Mechanism of solid waste generation.

the waste quantity generated by people, the composition of the waste, and its sources. Information collected periodically from such assessment will help in identifying equipment at the facility, the formulation of plans, and programs for the management system. Waste volume and composition vary from region to region and also depend on the lifestyle and income level of the population. The primary constituents in waste are decomposable organic matter and paper followed by inorganic material comprised of metals, plastics, textiles, glass, and so on. Due to the socioeconomic status of community, waste generation may vary, but the ultimate proportion reaching disposal sites in urban areas remains the same. The composition of waste varies according to the socio-economic status of the community, hence in developing countries a major component of waste is biodegradable and organic, whilst in developed countries, there is a larger component of inorganic waste.

- Composition of paper in solid waste ranges from 1% to 5% in low-income countries whereas, in high-income countries, it varies from 20% to 50%.
- Plastic composition in solid waste varies from 1% to 5% and 5% to 10% in low- and high-income countries, respectively.
- Ash and fine material fractions in solid waste ranges from 15% to 60% and 3% to 10% in low- and high-income countries, respectively.
- In low-income countries, moisture content found in solid waste is around 30%–40% and, in high-income countries, it is 15%–30% (Short Term Course on Solid Waste Management, CED 2012).

Variation in waste composition mainly occurs due to density, moisture content, and distribution of waste. Generally, the density of waste changes during transfer from source to disposal facility. During the transfer, many factors such as handling methods, storage of waste at the primary and secondary location, weather exposure, decomposition of waste, and so on affect the density.

2.1.2.1 Waste Characteristics

Waste characteristics are divided into two categories: physical and chemical. Waste characterization plays a vital role in identifying a suitable processing technology.

2.1.2.2 Physical Characteristics

Analysis of the physical characteristics of waste provides necessary information for the selection of waste handling equipment and also for designing processing methods and the ultimate disposal of waste. There are three principal components that determine the physical characteristics:

- Moisture content. This is defined as the *weight of water* (wet weight–dry weight) to total weight of wet waste (CED 2012). A great deal of moisture in waste affects various processing technologies that require dry waste. As the moisture content increases, the weight of the waste also increases and this causes problems during collection and transportation. In addition to this, moisture content determines the economic feasibility of mass burn or incineration technology. To remove excess moisture and to raise the temperature of water vapors, an excess amount of energy is required (Reeb and Milota 1999).

$$\text{Moisture content} = \frac{\text{Wet weight} - \text{Dry weight}}{\text{Wet weight}} \times 100$$

- Density of waste. Density is defined as mass per unit volume (kg/m^3). Density is the most significant physical characteristic of waste. Waste should be compacted to maximum density so that it occupies less area in a sanitary landfill. For up to 75% of waste, the volume can be reduced by any compaction equipment. This results in an increase in initial density of $100-400$ kg/m^3 (CED 2012). Density frequently varies as waste goes through the various processes from the source of generation to ultimate disposal and becomes affected by decomposition and changes in temperature.
- Size. Size of waste plays a significant role in identifying equipment such as separators and shredders that are suitable for the processing of particular sizes of waste. Size distribution analysis is performed in the same manner as for soil particles.

2.1.2.3 Chemical Characteristics

Chemical characteristics provide knowledge of various chemical elements of waste as well as heating value. The principal components of chemical waste include the following:

- Lipids. These are fats, grease, and oils that are generated from food waste, cooking oils, fats, and so on. The calorific value of lipids is about 38,000 kJ/kg, which makes them more suitable for energy recovery (CED 2012). They become liquid at ambient temperature and are biodegradable in nature but the rate of degradation slows because of the low solubility of lipids in water (CED 2012).
- Proteins. These consist of elements such as carbon, nitrogen, oxygen, hydrogen, and organic waste with amines. Proteins are generated from food and garden wastes. Protein containing the amine group produces an unpleasant odor on partial decomposition.
- Carbohydrates. A major portion of MSW is organic waste, consisting of food and yard waste, which contains carbohydrates in the form of polymers of sugar. Carbohydrates rapidly decompose when they come into contact with air and produce CO_2 (gas), H_2O (liquid), and CH_4 (gas).

2.1.2.4 Proximate Analysis

Proximate analysis is carried out to examine various parameters of combustible components of solid waste. These parameters include moisture content, volatile combustible matter, and fixed carbon and ash content.

- Moisture content is defined as the loss of the moisture in combustible waste when heated to 105°C for 1 h (Tchobanoglous et al. 1993).
- Volatile matter is the combustible solid waste that directly evaporates when heated to 950°C in a covered crucible for 7 min.
- Fixed carbon is the combustible waste that remains after removal of volatile substances, ash, and moisture content. It must be burned in a solid state (ASTM D3172, 2013).

$$\% \text{ Fixed carbon} = 100\% - \%\text{Moisture} - \%\text{Ash} - \%\text{Volatile matter}$$

2.1.2.5 Ultimate Analysis

The ultimate analysis of combustible solid waste includes an examination of percentage carbon, nitrogen, hydrogen, oxygen, sulfur, and ash. Halogens are also included in the analysis as they contribute to emission during the combustion process (ASTM D3172, 2013). Ultimate analysis is also performed to identify the chemical characteristics of organic waste in MSW (Tchobanoglous et al. 1993). Ultimate analysis includes an examination of cellulose, proteins, lignins, fats, hydrocarbon polymers, and inorganic content in MSW (Tchobanoglous et al. 1993; ASTM D3172, 2013). Ultimate analysis is also performed to examine waste mix suitable for achieving the required carbon-to-nitrogen (C/N) ratio for biological conversion processes.

The heating value is defined as the amount of heat generated during the combustion of solid waste under standard conditions. It is analyzed to determine the potential of a solid waste sample to be used as fuel in the incineration process. Plastic wastes have a high heat value, a low ash and low to moderate moisture content. Paper and cardboard have intermediate heating values due to average carbon content as well as a moderate moisture content; whereas food waste or green waste has a very low heating value due to a high moisture content (ASTM D3172, 2013). Various studies have shown that solid waste generated in high-income countries have higher heating values due to the presence of high carbon content in waste; whereas low-income countries or developing countries have little heat value because of high moisture content and low carbon content in waste (CED 2012).

2.2 MSWM: Functional System

The functional system of MSW involves the management of waste from the point of generation, waste handling, sorting, storage, and processing at source along with transfer, transport, and disposal. In the past, there was less concern about management practice but, in more recent times, there has been greater focus on this because of public health concerns and aesthetic considerations. The MSWM first gained attention in 500 BC. with very few techniques and means to manage solid waste. As time passed, people began to realize that this is one of the important sectors of waste management that has to be given major attention.

2.2.1 Historical Development of MSWM System

The poor state of solid waste management (SWM) in urban areas of developing countries is now not only an environmental problem but also a major social handicap (Daskalopoulos et al. 1998).

The generation of waste has been a part of history since the earliest times. With the unavailability of proper waste management systems, large amounts of waste were released back into groundwater, creating a significant environmental impact.

The root of waste management was planted in early 1751 by Corbyn Morris in London. He was concerned about people's health. He proposed that the health of the commoners was of great importance and firmly promoted his idea to have uniform public management. He arranged that the garbage and scraps of the city should be diverted to the River Thames in order to have a clean and sanitized city.

Current research shows that an organized SWM in the city came into place in the eighteenth century. This appeared 50 years before the Public Health Act. At that time, major portions of the waste were in the form of coal ash and this played a major role in brickmaking and soil improvement. This regulation opened the way for resource recovery and waste collection systems. Thus, waste collectors were encouraged to effectively collect with efficiency 100% of residual waste from the booming informal sector in the streets. Therefore, the dust yard system became an early example of a well-organized MSWM system. It worked successfully up to the mid-1850s until the market of "dust" that is, the market of coal ash, which had played a prominent role in brickmaking and soil improvement in the eighteenth century, came down. This MSW system laid an early foundation of waste management.

The middle of the nineteenth century was devastated by a sudden and disturbing cholera outbreak. Public health then became a major issue. In 1842, Edwin Chadwick, a social activist, published an eye-opening report named "The Sanitary Condition of the Labouring Population." In his influential report, he shed light on the improvement in people's health and the welfare of the city's population.

The Disease Prevention Act of 1846 started the process of regulation regarding waste management systems. First, the metropolitan board of works introduced the "citywide broadway concept." It centralized the sanitation regulations for the sudden and rapid expanding city. The Public Health Act 1875 introduced a regulation stating that every house would deposit its weekly waste in "movable receptacles." This was the first ever "dust bin" concept.

In Europe and North America in the twentieth century, similar types of MSWM developed in cities. In 1875, New York City became the first US city to have a public sector garbage management. During an earlier phase of waste management, the solid waste was loaded onto open trucks pulled by pairs of horses. In the early twentieth century, the first closed-body trucks were introduced and a dumping lever mechanism was introduced in 1920 in Britain. Subsequently, the garbage-carrying trucks were loaded using a hopper mechanism. In 1938, hydraulic compactors were introduced (http://central.gutenberg.org/articles/History_of_waste_management).

2.2.2 Evolution of MSWM System

Table 2.2 presents the evolution of the MSWM system. It indicates a time line of trash, its location, and management system.

2.2.3 Modern MSWM System Techniques in Developing Countries

MSWM is an integral part of every human society. The approach to MSWM should be in accordance with global trends and should be compatible with the environment. The financial status of the country determines whether or not the particular option selected for MSWM will be sustainable. High-income countries such as Japan and South Korea can spend more on the 4R technologies (reduce, recycle, recover, and reuse). The world, today, is focusing on the concept of zero waste and zero landfill. This goal is quite expensive for financially weak countries such as India or Indonesia to attain. Hence, there is a need to assess the effects of implementing MSWM systems in developing countries (Shekdar 2009).

It is clear that developing countries are far behind developed countries in terms of the practical implementation of the MSWM technologies available. Although cost-effective technologies are available for use in developing countries, the effective laws are, in many cases, not sufficiently stringent. Also, the economic condition of the government is so poor that it cannot afford to use the new technologies. The modern MSWM techniques for developing countries are presented in the following section.

TABLE 2.2

Evolution of Waste Management System

TimeLine of Trash		
Date	Location	Notes
6500 BC	North America	Archeological studies show that North America produced an average of 5.3 pounds of waste a day in 6500 BC and management systems to handle the enormous waste were not effective at that time.
1388 BC	England	As the amount of waste was increasing along with the population and urbanization, the English Parliament banned waste dispersal in public waterways.
500 BC	Athens, Greece	First organized municipal dump was set up during this era in Athens. Regulations were made to dump the waste at least a mile from the city limits.
1842	England	"Age of sanitation" begins during this era.
1874	Nottingham, England	A new technology called "the Destructor" was invented during this time which provided the first systematic incineration of refuse.
1885	Governor's Island, New York	The first garbage incinerator was built in New York.
1896	United States	Waste reduction plants arrived in the United States during this era and they were used for compressing organic waste. Later, they were closed because of noxious emissions.
1898	New York	New York opened its first waste sorting plant for recycling.
Turn of century: starting of 1900.		By the turn of the century, the garbage problem was seen as one of the greatest problems for local authorities.
1914	United States	During this period, there were about 300 incinerators in the United States for burning trash.
1920s		Landfill was becoming a popular way of reclaiming swamp land while removing trash.
1954	Olympia, Washington	The idea of recycling grew and this led to the development of a new strategy. Customers were paid for returning aluminum cans.
1965	United States	The first federal solid waste management laws were enacted.
1968		Companies began to buy back recycling containers.
1970	United States	The first Earth Day was celebrated and the Environmental Protection Agency (EPA) was created. It introduced the Resource Recovery Act.
1976	United States	The Resource Conservation and Recovery Act (RCRA) was implemented to emphasize recycling and household waste management. This was the result of two major events: the oil embargo and the discovery (or recognition) of Love Canal.
1979	United States	The EPA issued criteria prohibiting open dumping.

Source: Rotten Truth: About Garbage. A garbage timeline. Association of Science-Technology Centers Incorporated and the Smithsonian Institution Traveling Exhibition Service. www.astc.org/exhibitions/rotten/timeline.htm.

2.2.3.1 Techniques Available for Implementation of MSWM Systems in Developing Countries

With the advancement in developing countries, people are no less aware of the pollution created by MSW. The public in developing countries now understand the importance of implementing good management systems to combat the SWM problem. Concerns about waste management are now understood widely. Increasing apprehension can be seen in the level of demand for more viable technical assistance in the field of waste management. Simply borrowing advanced and expensive technology from developed countries cannot solve the problem of waste management in developing countries. The urgent need is to design an appropriate technology that is more functional in developing countries.

In some cases, the introduction of sophisticated, expensive techniques from developed countries has resulted in failure because they are unfit for the prevailing system of management. Matsufuji Yasushi of Fukuoka University lists "six M's."

- Machines
- Manpower
- Material
- Management
- Maintenance
- Motivation

These are very necessary to the improvement of SWM in developing countries. The major obstacles that hinder SWM are financial resources, manpower, and physical resources in the form of materials.

2.2.3.2 Status of MSWM Systems in Developing Countries together with Recent Available Technologies

- MSW collection is the responsibility of municipal corporations. Community bins in most cities placed at different points and these sometimes lead to the creation of unauthorized open collection points. House-to-house collection is being introduced in the megacities.
- With the help of nongovernmental organizations (NGOs), municipal corporations have teamed up to manage a huge quantity of waste in cities. Private contractors have also been brought in for secondary transportation of community bin waste or for transporting waste from open collection points to the disposal site (Wang 2014).
- At present, some of the NGOs and citizen's committees supervise segregation, collection, and disposal of waste. At some places, welfare associations on monthly payment arrange the collection of

waste. Sweepers are allotted a specific area (around 250 m²) (Colon and Fawcett 2006; Nema 2004; Malviya et al. 2002; Kansal et al. 1998; Bhide and Shekdar 1998).

- In developing countries, waste sometimes remains uncollected because the streets are both unplanned and overcrowded, and this causes discomfort to people. Most cities are unable to provide a specific waste collection system for all parts of the city.

- In developing countries, two innovative concepts of waste disposal are being implemented. These are composting and waste-to-energy (WtE). The composting method includes both aerobic and vermi-composting whereas WtE includes different treatment processes such as incineration, pelletization and bio-methanation (Sharholy 2007). Techniques such as WtE are tested with positive results in many developed countries. In some developing countries such as India, WtE is a relatively a new concept. Therefore, landfill is still the final destination for the disposal of waste (Fulekar et al. 2014).

- Due to inappropriate MSW management, 90% of waste is directly disposed of onto the land of various cities and towns. In many coastal areas, such dumping leads to leaching of heavy metals (Dev and Yedla 2015). The availability of land for waste disposal is very limited in developing countries (Mor et al. 2006; Siddiqui et al. 2006; Sharholy et al. 2006; Gupta 1998; Das et al. 2015; Kansal et al. 1998; Chakrabarty et al. 1995; Khan 1994). In most urban centers, MSW is disposed of in low-lying areas mostly outside the cities without following the ideology of sanitary landfill where waste is dumped at a particular site and allowed to degrade biologically, chemically, and physically in isolation from human settlement (Dev and Yedla 2015).

- Incineration is a thermal treatment option that reduces the volume of waste and, thus, saves the encroachment of waste onto land. It also lessens the extra environmental burden. This process is quite expensive due to complex technology. Incineration requires large-scale burning and it also causes air pollution, which adds extra work to manage. In developing countries such as India, incineration is not very practical as the waste contains high moisture content and high organic content and, also, the calorific value is quite low, that is, 800–1100 kcal/kg (Kanal 2008; Bhide and Shekdar 1998). Therefore, the developing countries with waste having the same characteristics might undergo anaerobic digestion and biomethanation. These two techniques are less costly and they are appropriate for the process of degrading high organic matter as a renewable resource.

- In developing countries, MSW is openly dumped and this is not advisable either for human health or for environmental care. Open dumping is found in 90% of India, 85% of Sri Lanka, 65% of Thailand, and 50% of China (Ojha et al. 2012). Appropriate technologies such as composting and WtE methods should be used in order to have a proper waste management system.

- The sustainability of any waste management technique depends on costs and the financial status of a region. An organized method of segregation and collection exists in developed countries as they are economically strong enough to purchase any costly technology but developing countries lack this financial foundation. Manual segregation and cheap technologies are still used in developing countries.

- Proper segregation of waste and recycling should be adhered to in developing countries as most of the valuable material that can again be used as a secondary resource is lost due to lack of segregation practice. Figure 2.2 shows the percentage of physical characteristics found in MSW in different countries.

- The aim of developing countries is to eliminate landfill in MSWM systems. These countries experienced rapid growth in the latter years of the twentieth century. Regular follow-up in the system provides reliable data for well-run operations in the waste management system and this lays the basis for technical training in these fields.

- Recycling is being increased in developed countries to reduce the burden on landfill. Extended producer responsibility (EPR) is being implemented to ensure the safe dumping of waste.

- Similarly, notable literature is available on diverse features of MSWM. Manual labor can be substituted by required techniques. Sustained planning and funding is made available so that different processing remains uninterrupted. The awareness level about waste management is quite high. The current global trend contributes toward sustainable ideology. National programs have been launched in Japan,

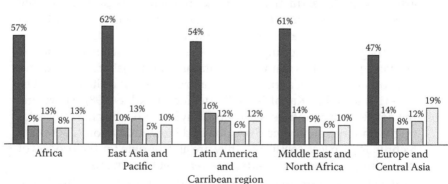

FIGURE 2.2
Percentage of physical characteristics found in MSW. (From World Bank, *What a Waste: A Global Review of Solid Waste Management*. Washington, DC: World Bank, 2012.)

TABLE 2.3

Waste Generation, Recycling, and Solid Waste Disposal in Japan, South Korea, and Taiwan

Country	National Program	Plan Period	Waste Generation	Recycling	Solid Waste Disposal
Japan	Establishing a sound material	2000–2010	Reduction by 20%	Increase by 40%	Reduction by 50%
South Korea	Firm establishment of a sustainable and resource-circulating socioeconomic foundation	2002–2010	Reduction by 12%	Increase by 15%	Reduction by 22%
Taiwan	Complete recycling for zero waste	Initiated in 2003	—	154 tons were recycled in 2007, 199 tons in 2011, and this is due to increase to 316 tons by 2020	No waste will be land filled in 2020

Source: Shekdar, V.A., *Waste Management*, 29, 1438–1448, 2009.

South Korea, and Taiwan. Time-launched and goal-driven programs have been launched (Shekdar 2009). Table 2.3 presents the national plan, period of the plan, waste generation, recycling, and solid waste disposal in different continents.

- The Integrated Municipal Solid Waste Management (IMSWM) system is an interrogative concept of waste prevention. It touches on all aspects of waste management such as recycling, composting, and dumping programs. A successful IMSWM system encompasses an authentic plan that covers all the compartments of waste management, including environmental and human health. This concept is widely practiced in developed countries.

- System analysis is also a very important technique for handling MSW through a wide range of comprehensive methodologies. It includes certain engineering models and assessment tools for solving the challenges and barriers involved in waste management.

2.2.3.3 Basic Comparison between Available Technologies Used in Developing and Developed Countries

Figure 2.3 shows the applicability of conventional and updated technologies used in the MSWM sector on all the continents. This figure describes the

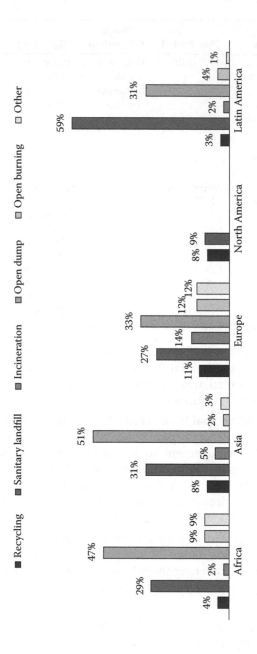

FIGURE 2.3
Applicability of the conventional and updated technologies used in the municipal solid waste management sectors in the named continents. (From World Bank, *What a Waste: A Global Review of Solid Waste Management.* Washington, DC: World Bank, 2012.)

recycling technique that is used only in developed countries and the conventional landfill methods that are still being followed in the developing countries. Only in a few megacities is the advanced technique being used within the developing countries. The Asian and African continents are still predominantly using the traditional open dumping method which means that MSWs are not being treated and this results in environmental pollution. This is in contrast to North America, where the open dumping method is being phased out completely. Incineration techniques are mostly being practiced in Europe and Asia and within high-income countries.

2.2.3.4 Problems to Be Faced while Implementing MSWM Systems in Developing Countries

There are several guidelines and procedures that need to be followed before starting any implementation techniques for MSWM systems. But if, in any sector, these guidelines are not being followed, then several problems might hinder the path of its implementation.

The countries that can best achieve environmental, social, economic, and technical goals are best at managing their implementation techniques. Implementing MSW management in developing countries is a little more problematic as they face several problems in different sectors. The basic problems faced in developing countries are enumerated in this section.

- Developing countries are facing rapid urbanization and population growth, and this sometimes exceeds the system threshold capacity, causing problems in managing the waste.
- Public health, environment, and waste management are interlinked. If proper waste management is not followed, it affects both public health and environment thus causing a major disruption in the waste management system.
- Waste management hierarchy is one of the biggest challenges and it needs a broad and comprehensive range of diverse treatment options such as composting and recycling to form a reliable infrastructure.
- Despite suitable planning models, insufficient information regarding segregation at source plays a major role in calculating the amount of material separated from the collected waste. The quantity of material segregated depends on the following factors with regard to coverage of collection system, separation efficiency of waste produce, and participation rate of waste producers.

Not much attention has been given to the following important factors in developing countries. In developing countries due to rapid urbanization and improving lifestyles, there are characteristic changes in the composition and quantity of waste. Municipal corporations are quite unable to complete the task. More than a quarter of waste remains uncollected (Pachuri and Batra 2001). Current findings show that unmanaged disposal of waste causes serious human health and environment concerns. This problem is quite well resolved by developing countries.

In practical terms, municipal governments tend to allocate low budgets to tackling waste. The public are less aware of waste management issues and so lower funds are made available, and this ultimately leads to further degradation in quality of service.

The revenue collected from the SWM is deposited in general municipal accounts with not much of it being spent on the particular purpose of waste management. The misallocation of waste funds is exacerbated when revenue is transferred to central government instead of local bodies. There is an absence of connection between revenue and the real level of service provision. In developing countries, due to lack of expertise and the availability of traditional waste management techniques, MSWM faces many problem. There is no universal remedy for the solid waste problem in developing countries.

Over 70% of the world's population lives in developing countries. Waste management in these countries is a major issue as the population is increasing at a faster rate than has ever been the case in industrialized countries (Olaosebikan et al. 2012) and waste management is an impending problem that needs to be solved.

TABLE 2.4

Problems Common to Waste Management in Developing Countries

External problems	• Population explosion, rapid urbanization, expansion of squatter settlements • Socioeconomic crisis • Lack of awareness in the field of public education and community participation
Simultaneously external and internal problems	• Large amount of municipal and industrial solid wastes, lack of waste-reduction efforts • Inefficient system of local authority • Negligence in solid waste problems among central and local government authorities • Lack of synchronization among sectors, organizations, and municipalities • Lack of comprehensible policy in informal sectors • Inadequate legal systems, insufficient law enforcement • Weak financial strata
Internal problems	• Lack of organizational capacity in MSWM • Lack of long-term and short-term planning • Insufficient operation and safeguarding structure for machinery and equipment • Lack of use of technology that is technically, economically, or socially adequate

Source: Yoshida, M. and Garg, V.K. *Waste Management Research*, 11, 142–151, 2000.

2.2.3.5 Categories of Problems Common to Waste Management in Developing Countries

Waste management in developing countries is hindered by different problems. These problems are further categorized into three categories:

- External problems
- Simultaneously external and internal problems
- Internal problems

The different problems faced by developing countries that hinder the smooth functioning of MSWM are presented in Table 2.4.

3

Generation of MSW

3.1 Source of Generation of MSW

Knowledge of the source of municipal solid waste (MSW) generation is vital to designing and operating the functional elements associated with the management of solid waste. To assess the current and future requirements for the budgeting, operating, and processing of MSW, it is important to collect information about the source and rate of waste generation in an area. The rate of solid waste generation varies from community to community, city to city, and country to country depending on various factors such as economic background, population density, industrialization, existing management status, and lifestyle (ADB 2011). The subsistence in MSW classifications has been creating confusion and making it difficult to interpret and compare the results of sources of generation. The sources of generation of MSW have been identified as coming from different classes such as residential, commercial, institutional, construction and demolition (Das 2014), agricultural, and animal husbandry, as shown in Figure 3.1. Different sources of waste are discussed separately as follows:

- *Household waste*: In most developing countries, the largest source of MSW generation is from households. It is also known as *domestic or residential waste* and it is generated from daily household consumption, including food waste, paper, plastic, metal and glass containers, rags, soiled papers, and so on, as shown in Figure 3.2. The kitchen waste in developing countries is almost two-thirds the total waste. Table 3.1 shows the data for household waste generation in five developing countries.
- *Commercial waste*: Markets are the most important source of commercial waste, consisting of biodegradable waste. Enormous quantities of plastics, papers, and cardboards are generated by general stores, as shown in Figure 3.3.
- *Institutional waste*: Institutions such as schools, government offices, and hospitals are increasing significantly in developing countries coupled with population growth. Paper predominates under the waste category in most institutions.

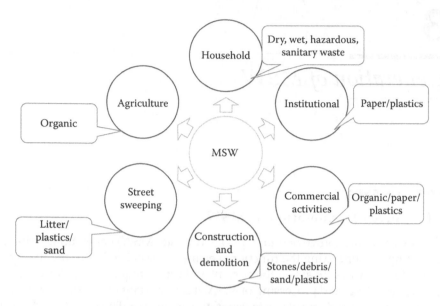

FIGURE 3.1
Schematic flow diagram of MSW generation from different sources in India.

FIGURE 3.2
Source of MSW generation from households.

TABLE 3.1

Household Waste Generation in Five Countries

Country	2005	2010	2015	2020
Japan	12,753	12,719	12,614	12,602
China	31,921	46,635	65,641	90,683
India	23,085	28,662	35,438	43,702
Korea	3,082	3,538	3,904	4,207
Malaysia	688	991	1,407	1,984

Source: Kanamori, Y. Household consumption change and household waste generation from household activities in Asian countries. *Berlin Conference on the Human Dimensions of Global Environmental Change*, Berlin, 2012.

FIGURE 3.3
Source of MSW generation from commercial areas.

- *Street sweeping*: This has been one of the largest sources of MSW generation and is a key facet of solid waste management (SWM) in developing countries. It consists of paper, plastic, litter, sand, stones, spilled loads, and debris from traffic accidents. The waste reaching the streets is due to lack of awareness among the population, littering by pedestrians, from vehicles, and by roadside dwellers. Public education in relation to participation is the key factor in MSW generation on the streets, as shown in Figure 3.4. Waste on the streets creates an adverse visual impact, particularly on visitors, and, thus, indirectly affects the economy of a city.

- *Construction and demolition (C&D) waste*: The trends toward urbanization and industrialization have led to the generation of C&D waste in many developing countries. This type of waste usually includes broken concrete, scrap and wood products, board, glass, old electrical material, tilling and related masonry, metal, and paints.

FIGURE 3.4
Source of MSW generation from street sweeping.

- *Agricultural waste*: Agricultural waste is one of the largest organic sources of waste generation in most developing countries. It has been found that organic waste from agricultural sources alone contributes to more than 350 MT/year (Asokan 2007). India generates 600 MT of waste from agriculture sources (Asokan 2007). The waste generated from agricultural sources includes waste from crops (wheat straw, bagasse, paddy, husk, vegetable waste, coconut husk, and shells) and other harvests (Sengupta 2002; Gupta 1998; Maudgal 2011). The rate of waste generation from the agricultural sector is dependent on the climatic conditions of the country as these are essential factors in crop production.

3.2 Factors Affecting Rate of Generation of MSW

Estimating the quantity and characteristics of MSW generation is a fundamental tool for the successful implementation of waste management options in a country. Waste generation depends on various factors such as population growth, economic status, culture, geographical pattern, and the level of commercial activity.

- *Population growth*: Among all other factors, population growth plays a particularly significant role in the rate of MSW generation. The increase in population in developing countries is coupled with the rise in urbanization. India's population is equivalent in size to the combined population of the United States, Indonesia, Brazil, Pakistan, Bangladesh, and Japan. Table 3.2. shows a significant increase of 50% in waste generation in one decade and this is expected to increase fivefold by 2041.

TABLE 3.2

Population and Waste Generation in India

Year	Population (Millions)	Per Capita/ kg/Year	Total Waste Generation Thousand Tons/ Year
2001	197.3	0.439	31.63
2011	260.1	0.498	47.3
2021	342.8	0.569	71.15
2031	451.8	0.649	107.01
2036	518.6	0.693	131.24
2041	595.4	0.741	160.96

Source: Annepu, R. K. Sustainable solid waste management in India. *Waste-To-Energy Research and Technology Council (WTERT) Bi-Annual Conference*, Columbia University, 2012.

In the past century, as the Indian population has grown and become urban affluent, waste production has risen tenfold. It has been found that approximately 45% of the country's population uses a proper sanitation infrastructure, while an average of 20% lacks adequate accessibility to water (Khatib 2011). Ultimately, the governments of developing countries are facing huge problems that are hindering the attainment of sustainable development and healthy municipal services.

- *Urbanization*: This is directly coupled with the rate of generation of MSW in developing countries. Rapid population growth, changing lifestyle, demand for a standard of living, and a tendency to relocate from stagnant and low-paying sectors of rural areas to higher-paying urban occupations has led to urbanization and, ultimately, a rise in the rate of waste generation. China, one of the fastest developing countries, has experienced high levels of urbanization with the urban population having increased from 58 million in 1949 to 670 million in 2010 (*China Statistical Year Book* 2011) and with ever-growing MSW generation. It is estimated that the solid waste generated in small, medium, and large cities and towns is about 0.1, 0.3–0.4, 0.5 (kg/c/d), respectively (Infrastructure Leasing and Financial Services 2010).

- *Industrialization*: The process of industrialization has transformed the entire infrastructure of developing countries as rural populations have been relocating to urban areas in search of sources of livelihoods. This has led to urbanization and, ultimately, to MSW generation. The average rate of waste generation has been found to be higher in industrialized cities such as in Japan where the relatively high per capita income is related to the rate of waste generation at 1.64 kg/c/d. In contrast, India, Bangladesh, and Sri Lanka, with low incomes have average rates of waste generation of around 0.64 kg/c/d (Dangi et al. 2011).

TABLE 3.3

Comparison of the Per Capita MSW Generation Rates in
Low-, Middle-, and High-Income Countries

Country	Per/Capita Urban MSW Generation (kg/day)	
	1999	2025
Low-income countries	0.45–0.9	0.6–1.0
Middle-income countries	0.52–1.1	0.8–1.5
High-income countries	1.1–5.07	1.1–4.5

Source: Annepu, R. K. Sustainable solid waste management in India. *Waste-To-Energy Research and Technology Council (WTERT) Bi-Annual Conference*, Columbia University, 2012.

- *Community activities*: MSW generation from community activities depends on lifestyle, culture, and economic background, as shown in Table 3.3. Depending on all these factors, it has been found that waste received from different communities shows vast differences in characteristics and composition. For example, some communities use and discard paper and plastics in vast amounts and others throw organic materials away (UNCHS 1989).

- *Economic level of different sectors*: The generation of MSW is commonly claimed to be coupled with economic activity and the level of production in different sectors of a society, as it is an inevitable consequence of production and consumption activities in any economy. The appropriate information relating to existing sectors is very important for waste characterization and evaluation and for selecting the appropriate waste management technology to fit with the existing conditions and requirements. The different sectors, based on economic levels in developing countries, are normally

  - High-income groups (HIG)
  - Middle-income groups (MIG)
  - Lower-income groups (LIG)
  - Slum areas

Cities with low and middle incomes have been found to generate a large proportion of organic waste, whereas cities with high incomes are relatively diversified with higher percentages of paper and plastics (UNEP 2011), as shown in Figure 3.5. Table 3.4 shows average MSW generation in Dhaka City, coupled with the economy and family size.

- *Attitude*: Due to people's lack of awareness about and interest in waste management, it has been a significant task to tackle the vast amounts of day-to-day waste generation. The problem is specifically

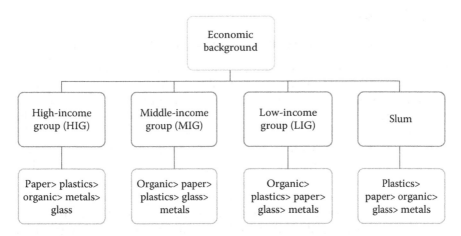

FIGURE 3.5
Schematic flow diagram of MSW generation in different economic sectors.

TABLE 3.4

MSW Generation Coupled with the Economy and Family Size in Dhaka City

Income Group	Family Size	Waste Generation (kg/day)
High-income group	3–5	0.50
Middle-income group	4–8	0.45
Lower-income group	4–9	0.29

Source: APO Report. *Solid Waste Management: Issues and Challenges in Asia, Survey on Solid Waste Management.* Tokyo, Japan: Asian Productivity Organization, 2004–2005.

in the developing countries where major factors include illiteracy, poverty, and irresponsible waste management. People tend to think that waste collection or management is the responsibility of the municipality and, thus, they are exempt from any responsibility. As a result, they have little concern for waste issues and so they tend not to make any effort to dispose of their waste in a proper place.

- *Seasonal variation*: This has a significant effect on the rate of MSW generation all over the world. The availability of fruits and vegetable and annual crops also affects MSW generation rate. For example, in summer, the rate of waste generation is influenced by the rapid degradation of a waste component.

3.3 MSW Generation Scenarios in Developing Countries

In developing countries, open dumping and open burning of MSW present a real threat to all living organisms as well as to the environment. MSW generation is increasing continuously in developing countries at a rate of 3%–7% per year (Dangi et al. 2011). The key issues of MSW generation in developing countries are caused by sizable population growth in urban centers, lack of legislation and policies for realistic long-term planning, inadequate storage, limited collection, lack of proper disposal, lack of appropriate technology and equipment, scavenging, lack of trained manpower, and so on. The majority of substances comprising MSW in developing countries (Walling et al. 2004) include paper, kitchen waste, plastics, metals, textiles, rubber, and glass (Getahun et al. 2012). In developing countries, the per capita generation rate ranges from 0.4 to 1.1 kg/d, reaching 2.4 kg/d in some urban areas and more in tourist areas (Hoonwerg and Giannelli 2007). It has been found that developing countries produce a lower level of waste per capita with a higher proportion of organic material in the MSW stream (UNEP 2005). At least 50% of waste in developing countries is assessed as organic in nature (biodegradable) (UNEP 2005). MSW services in developing countries are improving but still lag behind developed countries.

The use of municipal solid waste management (MSWM) in developing countries is expected to increase, but it is still in its early stages. To reduce the drastic disposal of MSW to landfill in developing countries, concerns about MSWM have become more fundamental. A 2012 World Bank report estimated that, by 2025, more than 40% of the world's MSW would be generated in the East Asia and Pacific region, and China, individually, will be producing around 180 million tons of MSW. The rate of MSW generation in developing countries is enormous and it is in direct proportion to the advancement and development of a nation. Failure to provide a management system results in greater environmental degradation with an increase in health risks to the urban population. The MSW generation of some developing countries is presented in Table 3.5. Brazil has developed its MSWM system over the past decade with the introduction of the National Policy of Solid Residues (NPSR) but it still lacks municipal services in major areas due to insufficient technical and financial resources. In Ghana, it is estimated that the average per capita daily waste generation rate is 0.92 kg/c/d and the average daily generation rate per average household of 4.27 persons is 3.93 kg/household/ day (Ansha 2014). The composition of MSW is mainly organic waste, plastics/rubber, paper/cardboard, ash/sand, textiles, nonferrous metal, glass/ceramics, ferrous metals, hazardous waste: 68.56%, 16.02%, 4.87%, 4.15%, 3.23%, 1.65%, 0.92%, 0.31%, and 0.29%, respectively. (Ansha 2014). Ghana has an efficient waste collection system, that is, a door-to-door collection system, which is common in most high-density areas (Agyepong 2011). The government of Ghana has adopted EPA (1999) and ensured the

TABLE 3.5

MSW Generation Data in Some Developing Countries

City	Year	Urban Population	MSW Generation (kg/capita/ day)	Total Tons/day
Albania: Tirana (UNSD, 2009)	2007	1,532,000	1.01	1,549
Belaru: Minsk (UNSD, 2009)	2007	1,806,200	1.21	2,182
Georgia: Batumi (UNSD, 2009)	2007	303,200	2.00	605
Georgia: Kutaisi (UNSD, 2009)	2007	185,960	3.06	568
Georgia: Tbilisi (UNSD, 2009)	2007	1,300,000	0.82	1,064
Benin: Parakou (UNSD, 2009)	2002	148,450	0.59	88
Burakina: Ouagadougou (UNSD, 2009)	2002	8,76,200	0.79	693
China: Hong Kong (UNSD, 2009)	2007	6,926,000	2.47	17,129
China: Macao (UNSD, 2009)	2007	525,760	1.51	793
Egypt: Cairo (UNSD, 2009)	2007	7,765,000	1.77	13,766
India: Annepu (R.K 2012)	2012	1,210,193,422	0.5	130,000
Indonesia: Jakarta (UNSD, 2009)	2005	8,962,000	0.88	7,896
Iran: Tehran (Damghani et al., 2008)	2005	8,203,666	0.88	7,044
Iraq: Baghdad (UNSD 2009)	2005	6,784,000	1.71	11,621
Ghana: Kumasi (ASASE, 2009)	2006	1,610,867	0.60	967
Guinea: Conakry (UNSD, 2009)	2007	3,000,000	0.24	725
Philippines: Manila (UNSD, 2009)	2007	1,660,714	3.00	4975
Philippines: Quezon City (UNSD, 2009)	2005	2,392,701	1.56	3,729
Niger: Zinder (UNSD, 2009)	2006	242,800	0.29	69
Zambia: Kusuka (UNSD, 2009)	2005	1,300,000	0.90	1,172
Zimbabwe: Hasare (UNSD, 2009)	2005	2,500,000	0.08	208

Source: World Bank, *What a Waste: A Global Review of Solid Waste Management.* Washington, DC: World Bank, 2012.

various stakeholders, including the private sector, the sustainable collection, disposal, and treatment of waste. It is also improving planning, monitoring, and the enforcement of appropriate regulations at the local level (Ansha 2014). However, Ghana is lacking stringent laws for the management of solid waste. In Pakistan, the government made clear the SWM rules under the local administration ordinance in 2001, but waste management in the province is still not fully developed and there is a need to provide more detailed guidelines in the country.

According to a study by APO (2005), Sri Lanka is facing MSWM problems due to the climatic condition of the country. Ragpickers are exposed to various diseases, especially skin disease, as they are moving from street to street and going to dump yards in search of valuable materials (APO Report 2005).

4

Management Practices of MSW in Developing Countries

4.1 Introduction

Rapid urbanization, population growth, and changes in lifestyles in developing countries have contributed to increased per capita municipal waste generation (Agdag 2008). In order to manage the vast quantities of MSW, the following are common practices used in different developing countries:

1. *Collection of waste*: MSW is collected either by mechanical or manual methods in developing countries Jafari et al. 2010. Collection of the waste is quite complex, as MSW is not only the waste coming from households but also from unspecified places. The waste is stored in bins provided. Waste is collected door to door or by using motorized vehicles that come at particular times. Municipalities are fully responsible for the collection of waste. They have to collect the waste either through their own infrastructures or through private sector contracts.

2. *Transport and transfer*: Refuse vehicles are used to transport MSW. The predominant vehicles used in transportation are compactors, tippers, dumpers, and stationary compactors. The loading and unloading of refuse vehicles is done either manually by sweepers and sanitary workers or through mechanized methods. In some cities or towns, transfer stations are present, which support the intermediate transfer of waste from the surrounding area.

3. *Processing and dumping*: The collected waste is either processed for energy or dumped in a landfill.

Annually, the daily generation rate of MSW is 2.4% in Pakistan. Proper disposal is an issue of concern. The waste is sometimes disposed of in water bodies and low-lying areas (Mahar et al. 2007). Treatment technologies such as incineration and sanitary landfilling are comparatively new in Pakistan (Veenstra 1997); unsophisticated open dumping is the most common practice (Mahar et al. 2007). In Sri Lanka, among all other types of solid waste, MSW is a growing concern. In urban areas, there are two types of councils: *municipal councils* (MCs) and *urban councils* (UCs); altogether there are 23 MCs and 41 UCs (Karunathne 2015). In Sri Lanka, over 6400 tons/day of MSW is generated Vidanaarachchi et al. 2006. The most common practice to manage waste is open burning and unscientific landfilling (Karunarathne 2015). Nepal is a hilly area, and the management of waste is quite a tedious task. In another report by Timilsina (2001), the average waste generated is 540 m³/day, with 72% of the total waste generated destined for final disposal in landfill.

4.2 Legislation and Laws in MSW in India

In view of the serious environmental degradation resulting from the unscientific disposal of MSW, the Ministry of Environment and Forests (MoEF), Government of India, drew up the Municipal Solid Waste (Management and Handling) Rules (2000); revisions are in process making it mandatory for urban local bodies (ULBs) to improve their solid waste management systems as envisaged in the rules. These rules dictate the procedures for waste collection, segregation, storage, transportation, processing, and disposal. Further, the rules mandate that all cities must set up suitable waste treatment and disposal facilities. These rules also specify standards for compost quality and health control. Compliance criteria for each and every stage of waste management—that is, collection, segregation at source, transportation, processing, and final disposal—are also prescribed in the MSW rules. The most important guidelines of the MSW rules are given in Table 4.1.

4.2.1 New MSW Draft Rule 2015

1. Storage of segregated solid waste at source
    a. Littering and open burning of solid waste shall be prohibited by all Urban Local Bodies within the area covered under their jurisdiction within six months from the date of the notification of the rule.
    b. To facilitate compliance, the following steps shall be taken by the ULB, namely:
        i. Create public awareness on:

TABLE 4.1

Important Guidelines of the MSW Rules

A	Parameters	Compliance Criteria
1	Collection of MSWs	1. Littering of MSW shall be prohibited in cities, towns, and in urban areas notified by the state governments. To prohibit littering and facilitate compliance, the following steps shall be taken by the municipal authority, namely:
		a. Organizing house-to-house collection of MSWs through any of the methods, like community bin collection (central bin), house-to-house collection, collection on regular pre-informed timings and scheduling by using bell ringing of musical vehicle (without exceeding permissible noise levels).
		b. Devising collection of waste from slums and squatter areas or localities, including hotels, restaurants, office complexes, and commercial areas.
		c. Wastes from slaughter houses, meat and fish markets, fruits and vegetable markets, which are biodegradable in nature, shall be managed to make use of such wastes.
		d. Biomedical wastes and industrial wastes shall not be mixed with MSWs and such wastes shall follow the rules separately specified for the purpose.
		e. Collected waste from residential and other areas shall be transferred to community bin by hand-driven containerized carts or other small vehicles.
		f. Horticultural and construction or demolition wastes or debris shall be separately collected and disposed off following proper norms. Similarly, wastes generated at dairies shall be regulated in accordance with the State laws.
		g. Waste (garbage, dry leaves) shall not be burnt.
		h. Stray animals shall not be allowed to move around waste storage facilities or at any other place in the city or town and shall be managed in accordance with the State laws.
		2. The municipal authority shall notify waste collection schedule and the likely method to be adopted for public benefit in a city or town.
		3. It shall be the responsibility of generator of wastes to avoid littering and ensure delivery of wastes in accordance with the collection and segregation system to be notified by the municipal authority as per para 1(2) of this Schedule.
2	Segregation of MSWs	In order to encourage the citizens, municipal authority shall organize awareness programmes for segregation of wastes and shall promote recycling or reuse of segregated materials. The municipal authority shall undertake phased programme to ensure community participation in waste segregation. For this purpose, regular meetings at quarterly intervals shall be arranged by the municipal authorities with representatives of local resident welfare associations and non-governmental organizations.

(Continued)

TABLE 4.1 (CONTINUED)

Important Guidelines of the MSW Rules

A	Parameters	Compliance Criteria
3	Storage of MSWs	Municipal authorities shall establish and maintain storage facilities in such a manner as they do not create unhygienic and insanitary conditions around it. Following criteria shall be taken into account while establishing and maintaining storage facilities, namely:
		1. Storage facilities shall be created and established by taking into account quantities of waste generation in a given area and the population densities. A storage facility shall be so placed that it is accessible to users.
		2. Storage facilities to be set up by municipal authorities or any other agency shall be so designed that wastes stored are not exposed to open atmosphere and shall be aesthetically acceptable and user-friendly.
		3. Storage facilities or bins shall have easy to operate design for handling, transfer and transportation of waste. Bins for storage of bio-degradable wastes shall be painted green, those for storage of recyclable wastes shall be printed white and those for storage of other wastes shall be printed black.
		4. Manual handling of waste shall be prohibited. If unavoidable due to constraints, manual handling shall be carried out under proper precaution with due care for safety of workers.
4	Transportation of MSWs	Vehicles used for transportation of wastes shall be covered. Waste should not be visible to public, nor exposed to open environment preventing their scattering. The following criteria shall be met, namely:
		1. The storage facilities set up by municipal authorities shall be daily attended for clearing of wastes. The bins or containers wherever placed shall be cleaned before they start overflowing.
		2. Transportation vehicles shall be so designed that multiple handling of wastes, prior to final disposal, is avoided.
5	Processing of MSWs	Municipal authorities shall adopt suitable technology or combination of such technologies to make use of wastes so as to minimize burden on landfill. Following criteria shall be adopted, namely:
		1. The biodegradable wastes shall be processed by composting, vermicomposting, anaerobic digestion, or any other appropriate biological processing for stabilization of wastes. It shall be ensured that compost or any other end product shall comply with standards as specified in Schedule-IV (http://www.moef.nic.in/legis/hsm/mswmhr.html). Mixed waste containing recoverable resources shall follow the route of recycling. Incineration with or without energy recovery including pelletisation can also be used for processing wastes in specific cases. Municipal authority or the operator of a facility wishing to use other state-of-the-art technologies shall approach the Central Pollution Control Board to get the standards laid down before applying for grant of authorization.

TABLE 4.1 (CONTINUED)

Important Guidelines of the MSW Rules

A	Parameters	Compliance Criteria
6	Disposal of MSWs	Land filling shall be restricted to non-biodegradable, inert waste and other waste that are not suitable either for recycling or for biological processing. Land filling shall also be carried out for residues of waste processing facilities as well as pre-processing rejects from waste processing facilities. Land filling of mixed waste shall be avoided unless the same is found unsuitable for waste processing. Under unavoidable circumstances or till installation of alternate facilities, land-filling shall be done following proper norms. Landfill sites shall meet the specifications as given in Schedule III.

Source: Ministry of Environment and Forests, Government of India. Notification. New Delhi, India. 2000. http://www.moef.nic.in/legis/hsm/mswmhr.html.

A. Reducing the generation of waste

B. Reusing the waste material to the extent possible

C. Processing food waste through home composting or community composting

D. Separately store bio-degradable wastes or wet waste and non bio-degradable including recyclable and combustible wastes or dry waste

E. Encouraging waste pickers to take away segregated recyclable material stored at source

F. Wrapping securely sanitary napkins/pads, tampons, infant and adult diapers, condoms, and menstrual cups before putting in domestic bin meant for non bio-degradable waste

G. Storing separately domestic hazardous wastes such as contaminated paint drums, pesticide cans, compact florescent lamps, tube lights, used Ni.cd batteries, used needles and syringes and health care waste

H. Storing separately construction and demolition waste at the source of waste generation

ii. Mandate citizens to store segregated wastes at source in separate domestic or trade bins and hand over these wastes separately to designated waste collectors for recycling, processing and disposal of solid waste.

2. Collection of solid wastes

a. Organize door-to-door collection of segregated bio-degradable or wet and non bio-degradable or dry solid wastes on a daily basis at pre-informed timings from all residential and non-residential premises including slums and informal settlements using motorized vehicles or containerized tricycles, handcarts or any other device which is suitable for collection of segregated waste without necessitating deposition of waste on the ground and multiple handling of waste.

b. Bio-degradable wastes from fruits and vegetable markets, meat and fish markets, horticulture waste from parks and gardens, shall be collected separately and to the extent feasible market waste may be processed or treated within the market area and horticulture waste within parks and gardens to make optimum use of such wastes and minimize the cost of collection and transportation of such waste.

c. Large institutional premises, residential complexes shall be motivated and incentivized to process bio-degradable waste within their campus to the extent it is feasible to do so.

d. Construction and demolition wastes or debris shall be separately collected and processed by the ULB or agency appointed by it for the purpose of its processing and disposal without mixing the same with bio-degradable, recyclable or non-recyclable combustible wastes that shall be collected from the door step.

e. Dairy waste shall be collected separately and regulated as may be prescribed in the municipal bye-laws.

f. Appropriate user fees or charges shall be levied from the waste generator for sustainability of operations of solid waste management.

3. Sweeping of street and cleaning of surface drains

a. ULB shall arrange for cleaning of roads, streets, lanes, bye lanes, surface drains, and public places at regular intervals and use containerized tricycles, containerized handcarts, and suitable motorized or non motorized devices for collection of such waste.

b. Synchronize with the system of secondary storage and transportation of such waste without necessitating deposition of such waste on the ground.

c. The waste shall not be mixed at any stage with the solid waste collected from the door step.

4. Secondary storage

a. Segregated solid waste collected from the door step shall, as far as practicable, be transported directly to respective waste 15 processing facility having facility of sorting and recovery of recyclable waste and in absence of such arrangement, the waste collected from the doorstep shall be taken to waste storage depots for secondary storage of waste.

b. Waste depots shall have covered containers for separate storage of bio-degradable or wet waste and non bio-degradable or dry waste collected from the doorstep.

c. The street sweepings and silt collected from the surface drains shall not be left or accumulated on roadsides and shall be transported directly to waste disposal facility or shall be temporarily stored in covered bins or containers kept separately for secondary storage of inert wastes at suitable locations for facilitating onward transportation of such waste to the disposal site; if the street sweepings contain bio-degradable or recyclable waste, such waste shall be segregated and sent to respective processing facility.

d. The secondary storage vehicles or containers shall synchronize with transportation system to avoid multiple handling of waste.

e. Secondary storage of waste in open spaces on the roadsides or open plots or in cylindrical concrete bins or open masonry bins shall be dispensed with.

f. ULBs shall, where necessary, establish and maintain covered secondary storage facilities in such a manner as they do not create unhygienic and insanitary conditions around it and the following criteria shall be taken into account while establishing and maintaining storage facilities:

   i. Storage facilities shall be created and established by taking into account quantities of waste generation in a given area and distance required to be travelled by the waste collectors to deposit the waste at the storage facilities.

   iii. Storage facility shall be so placed that it is accessible to users.

   iv. Storage facilities to be set up by ULBs or any other agency shall be so designed that waste stored is not exposed to open atmosphere and shall be aesthetically acceptable and user-friendly and shall not be accessible to stray animals and birds.

   v. Storage facilities shall be a covered bins or containers of appropriate design including flaps and shall have "easy to operate" design for handling, transfer and transportation of waste and handling during evacuation of waste should be user friendly and not cumbersome.

   vi. Bins for storage of bio-degradable wastes shall be painted green, those for storage of recyclable wastes shall be painted blue, and those for storage of street sweepings and silt shall be painted black.

    vii. The design shall be developed in accordance with local practices and materials available to ensure minimal impact on health and environment.

    vii. Manual handling of waste shall be minimized, and waste handlers shall be given personal protection equipment to avoid direct contact with the waste.

    vii. Construction and demolition waste shall be separately stored in enclosed areas or containers separately without mixing these waste with waste collected from door step or street sweepings.

    ix. Bio-medical wastes, industrial wastes, e-waste and domestic hazardous wastes shall not be brought to the secondary waste storage depots or mixed with solid wastes and such wastes shall be handled as specified in specific rules framed for management of such wastes and domestic hazardous waste may be handled as directed by the state pollution control board or pollution control committee.

    x. Secondary storage bins if placed shall be cleaned at regular intervals at least once in a month and shall be painted at least once in a year.

5. Material recovery facilities

    The ULB shall designate temporary storage spaces and setup material recovery facility where non bio-degradable or recyclable solid waste collected from the doorstep shall be temporarily stored by the ULB or operator of the facility before solid waste processing or disposal is taken up in order to facilitate segregation, sorting and recovery of various components of recyclable waste by informal sector of waste pickers or any other staff or agency engaged by the ULB for the purpose and such sorting facilities shall be so designed that the solid waste stored is not exposed to open atmosphere and shall be user-friendly.

6. Transportation of solid wastes

    a. Waste collected from the door step in motorized vehicles shall be directly transported to the processing facility through material recovery facility to be set up at the waste processing site or to the transfer station or transfer point or waste storage depots for facilitating, sorting and bulk transfer of waste to the processing facility in large hauling vehicles or containers.

    b. Vehicles used for transportation of wastes shall be covered and shall have a facility to prevent waste spillage and leachate dropping from the vehicles on the ground en-route to the processing or disposal facility.

c. Waste shall not be visible to public, nor exposed to open environment preventing their scattering.

d. Waste stored at the secondary waste storage depots in covered bins or containers shall be attended daily and waste picked up before container start overflowing.

e. Bio-degradable waste stored in green and recyclable and combustible, and domestic inert waste stored in blue containers at the waste storage depots shall be transported to respective processing facilities in a segregated manner and the inerts street sweepings and silt collected from the drains shall be stored in black containers and shall not be allowed to be mixed with the waste collected from the door step or those stored in green or blue containers and such inert waste shall be directly taken to waste disposal facility or to the processing facility, if and when created for processing.

f. Separate transportation of domestic hazardous waste shall be arranged as directed by the State Pollution Control Board or the pollution control committee, as the case may be.

g. Construction and demolition waste shall be transported in covered vehicles separately to construction and demolition waste processing facility; and transportation vehicles shall be covered and so designed that multiple handling of wastes, prior to final disposal, is avoided.

7. Processing of solid wastes

a. Urban local bodies shall adopt suitable technology or combination of appropriate technologies, with emphasis on decentralized processing, to make use of all components of wastes that can be processed so as to minimize burden on landfill. Following criteria shall be adopted:

b. Biodegradable wastes shall be processed by bio-methanation, composting, vermi-composting, anaerobic digestion or any other appropriate biological processing for stabilization of wastes.

c. It shall be ensured that composting or any other end product shall comply with standards as specified in Schedule-II and also ensure that no damage is caused to the environment during this process.

d. To the extent feasible market waste may be processed or treated within the market area and horticulture waste within parks and gardens to make optimum use of such wastes and minimize the cost of collection and transportation of such waste.

e. Dairy waste shall be used for bio-methanation or vermi-composting or aerobic composting, either separately or with other bio-degradable solid waste.

f. Arrangement shall be made to provide segregated recyclable material to the recycling industry through waste pickers or any other agency engaged or authorized by the urban local body for the purpose.

g. Residual combustible wastes shall be utilized for supplying as a feedstock for preparing refuse derived fuel (RDF) or for generating energy or power from the waste by adopting proven waste to energy technologies for which emission standards as well as standards for dioxins and furans have been prescribed by the Central Pollution Control Board.

h. Non-recyclable plastics and other high calorific content waste may be utilized for co-processing in cement kilns or for polymer or fuel production or manufacturing of products such as door panels and the like nature.

i. Construction and demolition and other inert wastes shall be utilized for making bricks, pavement blocks, construction materials such as aggregates.

j. Urban local body or the operator of a facility planning to use other state-of-the-art technologies shall approach the Central Pollution Control Board to get the standards laid down before applying for grant of authorization.

8. Disposal of solid wastes

a. Land filling or dumping of mixed waste shall be stopped soon after the timeline as specified under for setting up and operation of sanitary landfill is over.

b. Landfill shall only be permitted for non-usable, non-recyclable, non-biodegradable, non-combustible and non-reactive inert waste and other wastes such as residues of waste processing facilities as well as preprocessing rejects from waste processing facilities and the landfill sites shall meet the specifications as given in Schedule-I, however every effort shall be made to recycle or reuse the rejects to achieve the desired objective of zero waste going to landfill.

c. Landfill site shall provide an appropriate facility for sorting, storing and transportation of recyclable material to the processing facility and ensure that such wastes do not get land filled.

d. All old open dumpsites and existing operational dumpsites shall be carefully investigated and analyzed about their potential of bio-mining and bio-remediation and actions shall be taken accordingly in cases where such course of action is found feasible.

e. In absence of potential of bio-mining and bio-remediation of dumpsite, it shall be scientifically capped as per landfill capping norms to prevent further damage to the environment.

(Source: http://www.moef.nic.in/sites/default/files/SWM%20Rules%20 2015%20-Vetted%201%20-%20final.pdf.)

4.3 Role of the Municipal Corporation

The municipal corporation is a very important body in the management of the civic and infrastructural needs of any particular region. It covers a vast network, including the environmental, sociocultural, technical, and legal sections of the waste management system.

The municipal corporation maintains the roads, streets, and flyovers, the water supply, open local spaces, sewage treatment and disposal, street cleanliness, and the cemeteries. It also deals with the modernization and upgradation of the existing system. For any municipal corporation, the collection and transportation of waste is a great matter of concern, and requires manpower and various transport vehicles, such as compactors, tippers, dumpers, stationary compactors, and so on.

To carry out the vast task of waste management, strong teamwork is required. Staff are deployed at various levels—that is, people from the level of sweepers and motor loaders up to supervisors and administrators. These people work day and night to manage and coordinate the functions of the municipal corporation.

The municipal corporation looks after the purchasing and standardization of community bins. It also looks after the deployment of nongovernmental organization (NGO) labor for collection points.

To make the system more efficient, municipal corporations are also becoming computerized, and the role of the municipal corporation is expanding into looking after computerized systems, including vehicle tracking and the implementation of radio frequency identification.

The municipal corporation also works toward the implementation of construction and demolition waste disposal, as per the debris management policy. It also aligns the workings of the MSWM department on functional lines. The municipal corporation also creates awareness among people by organizing different programs and street plays. It has a very important role in bringing changes to the organizational culture and introducing the latest technologies. The role of the municipal corporation is shown in Figure 4.1.

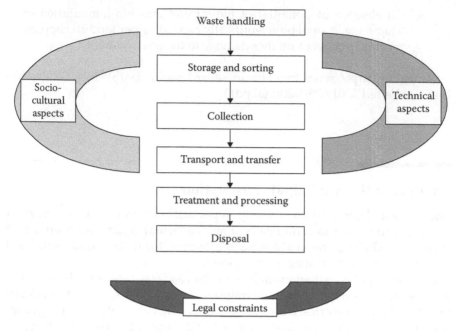

FIGURE 4.1
Role of the municipal corporation. (Modified from Sharholy et al., *Journal of Waste Management*, 27(4), 490–496, 2007.)

4.4 Role of Ragpickers in MSWM

Along with the official sector of MSWM, there is another unceremonious and unofficial sector consisting of people who work informally. The people of this sector are called *ragpickers*, who pick up rags and recover recyclable waste and other resources; they receive little recognition and are highly vulnerable. Their activities epitomize the informal sector, as this is labor-intensive, low-technology, low-pay, unrecorded, and unregulated work, often completed by individuals or family groups.

The job of the ragpickers is to sort and collect unused valuable waste, consisting of plastics, bottles, cardboard, tins, aluminum, iron, brass, and copper. After collection, these people sell the waste to scrap dealers. They provide primary collection and process collected materials into intermediate or final products using creativity and innovation to respond cost-effectively to market needs (Ahmed and Ali 2004).

In the trade of ragpicking, there are all types of people in groups irrespective of cast, creed, sex, or age, all belonging to the lowest economy. Informal waste recycling is carried out by poor and marginalized social groups, who resort to scavenging and waste picking for income generation and some even for everyday survival (Medina 2000).

There are three categories of ragpickers:

1. *Street waste pickers*: Collect the secondary raw materials recovered from mixed waste discarded on the street.

2. *Municipal waste pickers*: Collect waste from the vehicles that transport MSW to disposal sites.

3. *Dump site waste pickers*: Collect waste from dump sites. Dump sites contain waste from all across the city and are thus desirable places for waste pickers.

Informal recycling systems can bring significant economic benefit to developing countries. From a macroeconomic perspective, they are well adapted to the prevailing conditions: there is an abundant workforce, which earns very limited capital. They minimize capital expenditures and maximize hands (Haan et al.1998; Scheinberg 2001).

Though ragpickers contribute to a great extent in the minimization of waste and in sorting valuable secondary raw material, on a personal level they have poor living conditions and limited access to facilities and infrastructure. Ragpickers have a high risk of occupational health hazards, and they often suffer injuries and dog bites. The occupational health risks to waste pickers in developing countries are high because of their manual handling of materials and lack of protective clothing/equipment, resulting in direct contact with waste (Cointreau 2006).

5

Component Technologies for Municipal Solid Waste Management

Mukesh Kumar Awasthi, Amanullah Mahar,
Amjad Ali, Quan Wang, and Zengqiang Zhang

5.1 Introduction

Rapid urbanization is taking place in the main cities of developed and developing nations around the globe. This results in immense pressure on urban local bodies (ULBs) to manage their existing infrastructure and meet the needs of the growing population. The existing solid waste management infrastructures are insufficient to handle the ever-increasing amounts of mixed solid waste. Lack of public knowledge and weak institutional structure are the main reasons towards inefficient segregation, collection, transportation, treatment, and disposal of waste. Due to the weak collection and gathering systems, solid waste is littered on streets, which results in unhygienic conditions. Most of the solid waste treatment technologies have failed because of the poor quality of nonsegregated waste. Ultimately, a significant fraction of untreated solid waste reaches open dump sites in developing countries, which results in high levels of air, water, and soil pollution in nearby surroundings. To solve this growing problem in urban centers, ULBs should include efficient technological solutions to collect, transport, and treat solid waste.

This chapter mainly focuses on all the technical components of municipal solid waste (MSW) management. It includes house-to-house collection systems, details about transportation systems, and detailed information on various biological and thermal processing methods, such as composting, biomethanation, incineration, and refuse-derived fuel (RDF). The chapter also includes the various ultimate disposal methods and design criteria of sanitary solid waste landfills.

5.2 Collection of Municipal Solid Waste

The waste collection process is the transportation of MSW from the source of generation (residential, industrial, commercial, institutional) to the point of treatment or disposal. The primary purpose of MSW collection is not only to collect recyclable solid waste materials but also to transfer MSW from the source to a site where the loaded vehicle can be dumped. This site may be termed a *transfer station, landfill disposal site,* or *material processing facility.* This location needs to be far away from urban settlements to avoid the environmental impacts of MSW on urban dwellers. Every ULB or municipal corporation should inform residents of the waste collection schedule at regular intervals. The generator of solid waste should be responsible for the segregation and delivery of waste to authorized collectors as per the schedule of the municipal corporation. To inform citizens about proper waste segregation and collection, the municipal authority should regularly organize awareness drives in association with local communities and nongovernmental organizations (NGOs).

5.2.1 House-to-House Collection

In this type of waste collection system, waste collectors visit each house to collect garbage and users have to pay for it. The general plan of the residence plays a significant role in the frequency of collection as to whether they can accommodate dustbins or not; the use of wheelbarrows and brooms are required whether a dwelling or not; the use of wheelbarrows and brooms are required when a dwelling cannot store waste on the premises on a daily basis and it needs to be collected by municipal coporation workers. Dwellings with relatively large open areas rarely have any problem with the storage of waste in enclosed containers for short periods of time. This system is only acceptable if the following two conditions are met: (1) space should be available for sorting the container outdoors, and (2) containers should be equipped with well-fitting lids to prevent unpleasant odors from escaping and also to inhibit access by insects and other animals (Morrissey and Browne, 2004; Muzenda et al., 2012; Mrowczynska, 2014). Also, most residences require garbage disposal units for the disposal of kitchen wastes, because in small apartments there is little space for waste containers in the working area of the kitchen, imposing a maximum storage period of 24 h (Nema, 2004; Memon, 2010). Therefore, it is necessary to provide daily waste collection from each apartment or provide a decentralized communal waste container. Small shops and large markets where stalls are rented presently have a similar problem. Regular waste collection from each apartment is necessary for proper management, while market stalls may need a collection service three to four times a day (Table 5.1). In most developing countries such as India, the population density of most areas of major cities is

TABLE 5.1

Comparison of House-to-House Collection and Community Bin Collection

Types of Collection	Advantages	Disadvantages
Community bin collection	• Less cost-intensive than house-to-house collection • 24-hour availability to households	• Problems with illegal waste disposal because households find it inconvenient to carry their waste to the community bin • Resistance from neighbors ("not in my backyard" syndrome) • Nuisance from animals and vermin roaming the waste
House-to-house collection	• Convenience for households • Prevention of littering • Reduction of community bin • Segregated collection of waste	• Collection restricted to fixed collection times • Increased costs

Source: The Central Public Health and Environmental Engineering Organisation (CPHEEO), *Manual on Water Supply and Treatment*, 2000. http://cpheeo.nic.in/Watersupply.htm.

much higher than that in developed countries. Because external sites for waste storage are limited in these densely populated areas, the collection frequency should be four times per day (Mc-Dougall, 2001; Medina, 2010; Maudgal, 2011).

5.2.2 Community Bin System

This type of bin system is adopted in locations where community residents can easily access containers to throw away their waste. This system is comparatively affordable and widely adopted in developed and developing countries. To successfully implement this system, containers should be covered and regularly cleaned. Also, separate containers should be provided for recyclables, biodegradables, and paper. In the past, community bins have not been well equipped and properly received by citizens, and consequently most bins are not emptied on time and then they overflow, thereby creating unhygienic situations in the surrounding area. This problem is made worse because citizens tend to throw waste into the collection bins from a distance, because they dislike the odors coming from the containers. However, community bin collectors are very efficient for waste collection and they can still be used in selected situations. For instance, the number of waste containers can be increased and they can be placed in high-rise buildings, housing compounds, and small community areas. But frequent collection and cleaning are essential to avoid unpleasant odor due to anaerobic waste degradation and thereby enhance the acceptance of community bins. Meanwhile, community bins must be well designed to allow easy access for citizens. Ideally, bins will not be emptied but should be exchanged with well-equipped clean and empty containers by decentralized trucks. Otherwise, the containers can be unloaded from trucks mechanically, depending on what kind of advanced machinery is used by the municipal or city corporation. Table 5.1 provides comparative information on house-to-house collection and community bin collection systems.

5.3 Transportation

Waste handling and segregation are the main steps for final disposal. Handling involves the shifting of loaded containers to the site of collection. Waste segregation is a significant process for collection, handling, and proper storage of solid waste at the source level. Transportation involves two necessary steps: the waste is transferred from small collection vehicles to a large vehicle; the waste is then transported to a landfill site far away from the urban settlement for final disposal (Troschinetz et al., 2009; Khan, 1994; Khaitab, 2011) (Figure 5.1).

5.3.1 Technical Requirements of Municipal Solid Waste Transport Vehicles

The swift collection of MSW plays a significant role in solid waste management. This can be achieved through improved vehicle selection in urban as well as rural areas, taking into account waste quantities and characteristics, the condition of the roads, and the distance to the final dumping site. Tractor-trailers are most suitable for medium shipping distances less than 10 km to the final disposal site. Trucks are most appropriate for long distances more than 10 km due to the financial implications. Compactor trucks are most suitable for the collection of low-density waste less than 250 kg/m^3—that is, plastics, paper, packaging material—which may be the most common waste in urban areas and cities. Compactor trucks are not suitable for rural and small towns due to the high density of the waste—that is, dust, dirt, and organic waste (Pak EPA 2005).

FIGURE 5.1
Collection and transportation of municipal waste.

5.3.2 Types of Municipal Solid Waste Transportation Vehicles

The MSW collected from waste containers is transported to the final disposal sites using different types of vehicles. In towns and rural areas, tricycles or tractor-trailers are commonly utilized for the transportation of waste. Lorries and light motor vehicles are mainly used in large towns and cities. In metropolitan cities, trucks are used to transport waste from collection points to disposal sites (Pak EPA 2005).

5.3.3 Transfer Stations

Transfer stations are facilities where MSWs are unloaded from collection vehicles and briefly held; from there they are reloaded onto big vehicles for long-distance transportation to final disposal sites or landfills. Communities can save money on the labor and operation costs of shifting waste to faraway places by combining the loads of several individual waste collection trucks into a single load. Hence, there is a decrease in the total number of vehicles traveling to and from the disposal site. Though waste transfer stations have the advantage of reducing the impact of trucks going to and from the dump site, they can also cause an increase in traffic in proximity to where they are situated. Transfer stations should be carefully located, designed, and operated to avoid problems for dwellings nearby (Tchobanoglous and Kreith 2002) (Figure 5.2).

5.3.4 Optimization of Transportation Routes

An integrated MSW system is needed to develop a cost-effective and environmentally sustainable approach (Ionescu et al. 2013; Eriksson et al. 2014). Globally, municipal waste collection is recognized as accounting for the majority of expenditure on solid waste management. Low- and middle-income countries spend 80%–90% and 50%–80% of their waste management budgets, respectively, on the collection of waste (Weng and Fujiwara 2011). Therefore, MSW collection and transportation are issues that create

FIGURE 5.2
Waste collection vehicles in Sakon Nakhon.

difficulties in the development of an integrated waste management system (Consonni et al. 2011). Civil authorities are encouraged to develop sound strategies, especially in urban areas, to decrease the cost of transportation and collection (Massarutto et al. 2011). Therefore, the optimization of MSW collection and transport from source becomes an important aspect of waste management system design. In any city, waste sources are located at various places throughout the area in a heterogeneous way, causing an increase in waste collection and transportation costs (Das and Bhattacharyya 2015). Hence, a suitable waste collection and transportation approach can efficiently decrease waste gathering and transportation costs.

5.4 Biological and Thermal Processing Methods

5.4.1 Composting

Composting is the controlled microbial decomposition of the organic fraction of solid waste, under aerobic conditions, where microorganisms convert waste into a stable end product such as compost. The term *co-composting* is described as the composting of two or more substances together. In the composting process, the decomposition of the organic fraction of waste causes a reduction in its volume, weight, and moisture content, minimizes potential odor, decreases pathogens, and increases potential nutrients for agricultural application. The composting process may reduce the spread of disease because of the destruction of some pathogens and parasites at elevated temperatures. This process has been practiced for decades as a modern waste management alternative both in developed and developing countries. It diverts a significant portion of organic waste from municipal collection services and from final disposal sites, and therefore enhances the economic and environmental sustainability of waste management systems (Hsu and Lo 1999).

During composting, readily degradable organic waste stabilizes into carbon dioxide and water. Stabilization in the composting process depends on the end user's use of the by-product; also, users cannot achieve 100% stabilization, as this would destroy the soil-building properties of compost (Graves et al. 2010). During the thermophilic phase, high temperatures accelerate the breakdown of proteins, fats, and complex carbohydrates such as cellulose and hemicellulose. As these high-energy compounds become exhausted, the compost temperatures gradually decrease and mesophilic microorganisms once again overtake other types of microorganism during the final phase of maturation (Fourti et al. 2008). The rate of MSW decomposition and the activity of these microorganisms is encouraged through management of the carbon-to-nitrogen (C:N) ratio, pH, temperature, moisture content, and other nutrient levels, as well as by the composition of the starting materials.

A well-balanced composting process increases the rate of natural decomposition and generates sufficient heat to destroy weed seeds, pathogens, and fly larvae (Delgado-Moreno et al. 2009). Microorganisms play an important role in the decomposition of organic waste. Various species of unseen aerobic microorganisms decompose organic material as they grow and reproduce. Proper management of the composting process requires the nutrient balance, moisture content, oxygen supply, temperature, and pH levels to be maintained.

The nutrient balance in the compost mix is calculated by C:N. A C:N ratio between 20:1 to 40:1 is required for rapid composting (Graves et al. 2010). The ideal moisture content needed for the solid waste sample to prepare a quality compost mix is around 60% after mixing the ingredients (Graves et al. 2010). The proper biological process requires an adequate amount of oxygen, as oxygen affects the temperature, moisture, and carbon dioxide content in the compost pile. An increase in the temperature of the compost pile indicates the presence of active microbes and the breaking down of complex organic matter into a simpler form. The optimum range of pH for composting is 6.0–7.5 (Graves et al. 2010). To increase the rate of the composting process, it is necessary to balance and manage all these components.

While composting, waste passes through two periods: the first is active composting and the second is curing. During the active period, readily degradable waste is broken down into simple matter by individual aerobic microorganisms, whereas in the curing period, degraded material further breaks down into a simpler form, and at the end of period stabilized compost forms (Figure 5.3). The following is the oxidation process that occurs during composting:

$$\text{Organic materials} + O_2 + \text{microorganisms}$$
$$\rightarrow CO_2 + H_2O + NH_3 + PO_4^{2-} + SO_4^{2-} + \text{energy}$$

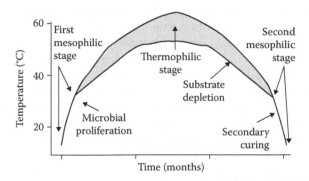

FIGURE 5.3
Different stages of the composting process. (Modified from Lekasi et al. 2003. Cattle manure quality in Maragua District, Central Kenya: Effect of management practices and development of simple methods of assessment. *Agricultural Ecosystems and Environment*, 94, 289–298.)

Three different temperature variations occur during the active composting process. These temperature ranges are defined by the presence of active microorganisms in that range. These temperature ranges are psychrophilic (below 50°F), mesophilic (50°F–105°F), and thermophilic (above 105°F) (Graves et al. 2010). In this period, carbon dioxide (CO_2) and ammonia (NH_3) are produced by the degradation of the organic fraction in the presence of oxygen, and pH also decreases due to the production of organic acids. This process makes pathogenic compost microbe free, as an increase in temperature creates unsuitable conditions for all the pathogenic bacteria. In the final stage, the temperature remains around 77°F–86°F. This indicates the compost has become stabilized and humification has occurred (Bundela et al. 2010). Composting of the organic fraction of MSW has become a vital alternative for MSW treatment, as the humic characteristics of compost optimize plant growth.

5.4.1.1 Chemical Transformations during Composting

During the chemical changes in composting, complex matter breaks down into a simpler form in the presence of various microorganisms. Before the microorganisms can synthesize new cellular material, they require sufficient energy for these processes. Two energy pathways followed by these heterotrophic microorganisms are respiration and fermentation. Respiration can be either aerobic or anaerobic (Alburquerque et al. 2009). Aerobic respiration is preferable to anaerobic respiration because of overall efficiency, as it produces a high amount of energy and raises the temperature during the process. Also, as it is more efficient than anaerobic respiration, it doesn't generate much odor. During aerobic composting, a large amount of energy is liberated due to the production of carbon dioxide and water from carbonaceous matter. This process takes place via a series of reactions wherein each intermediate compound acts as the starting point for other reactions (Graves et al. 2010). In anaerobic respiration, the microorganisms use electron acceptors other than oxygen to obtain energy. These electron acceptors include nitrates (NO_3), sulfates (SO_4), and carbonates (CO_3). Their use of these alternate electron acceptors in the energy-yielding metabolism produces unpleasant gases such as hydrogen sulfide (H_2S) and methane (CH_4) (Graves et al. 2010). Organic acid intermediates formed during anaerobic respiration affect the working of aerobic microorganisms, whereas in aerobic respiration these intermediates are readily consumed in subsequent reactions and don't produce odor as in anaerobic respiration (Graves et al. 2010).

5.4.1.2 Design of Compost Mixtures

Normal deterioration occurs in any heap of waste material regardless of whether the C:N proportion, dampness, and air circulation are outside the parameters for fertilizing the soil. Deterioration continues to ease back waste at a rate that is promptly perceptible and might likewise create foul smells. Fertilizing the soil is a guided push to expand the rate of general disintegration, diminish the generation of scents, and demolish pathogens, weed seeds, and fly hatchlings. The fertilizer blend must be intended to streamline conditions inside of the heap in terms of nourishment, the presence of oxygen and dampness, and pH and temperature levels so that a high rate of microbial movement is accomplished.

5.4.1.3 Components of Compost Mix

There are three major elements of a compost mix: the primary substrate, the amendment, and the bulking agent. The primary substrate is the waste that undergoes the composting process, whereas the amendment is the material that is added to balance the pH and the C:N ratio and to increase the moisture content of the waste sample. To maintain the structure and porosity of the compost pile, bulking agents are mixed in with the material (Nakasaki et al. 2009). With the help of these components, the physical and nutritional characteristics of the compost pile are maintained to create the perfect conditions for optimum microbial activity. The C:N ratio and moisture content are the two most crucial characteristics of a compost mix that need to be maintained in order to have rapid composting. If any one of the characteristics is not fulfilled, then an amendment is added. If the primary substrate's structure and porosity are not properly maintained, then bulking agents such as wood chips can be added to it (Albiach et al. 2000). There are some instances where neither the C:N ratio nor the moisture content of the compost mix can be balanced; amendments are then added so that either of them can come close to the ideal range or within the recommended range.

Moisture content is always preferred over C:N ratio if both of the characteristics cannot be balanced, as inappropriate moisture content severely affects the composting operation, more so than an unbalanced C:N ratio. After balancing the moisture content, the C:N ratio can be adjusted to a higher value. Higher C:N values will only slow the process, whereas lower C:N values generate odor and cause the loss of essential nutrients.

5.4.1.4 Types of Feedstock

Feedstocks are the materials that reach the composting facility for the production of compost mix. For the better operation of any composting facility, an efficient feedstock management plan is required. Improper management at the facility always results in odor concerns and a poor quality of compost at the end. An efficient material selection and management program will help the facility operator to gain the benefits that each feedstock can contribute to the compost mix. Some criteria and parameters have been proposed for testing compost stability and maturity, all of which express these characteristics as a function of composting time independently of

TABLE 5.2

Common Feed Stocks and Their Characteristics for the Rapid Composting Process

Feedstocks	Moisture Content	C:N Ratio	Bulk Density
High Carbon Content			
Hay	8–10	15–30	—
Corn stalks	12	60–70	32
Straw	5–20	40–150	50–400
Corn silage	65–68	40	—
Fall leaves	—	30–80	100–300
Sawdust	20–60	200–700	350–450
Brush, wood chips	—	100–500	—
Bark (paper mill waste)	—	100–130	—
Newspaper	3–8	400–800	200–250
Cardboard	8	500	250
Mixed paper	—	150–200	—
High Nitrogen Content			
Dairy manure	80	5–25	1,400
Poultry manure	20–40	5–15	1,500
Horse manure	65–80	10–20	—
Cull potatoes	70–80	18	1,500
Vegetable wastes	60–65	10–20	—
Coffee grounds	14–18	20	—
Grass clippings	8–10	15–25	—
Sewage sludge	—	9–25	—

Source: Rynk et al., *On-Farm Composting Handbook*, pp. 6–13, 106–113. Ithaca, NY: Northeast Regional Agricultural Engineering Service, 1992; Wong et al., *Bioresource Technology*, 100, 3324–3331, 2009.

the composting process or feedstock composition. On that basis, five major types of feedstocks are currently used for composting (Rynk 1992, and Wong et al. 2009) (Table 5.2).

MSW contains a heterogeneous organic fraction of feedstock material with widely varying sizes, shapes, and compositions. This can be difficult to directly use for composting and can lead to variable composting behavior. MSW received at a facility needs to undergo minimum size reduction and sorting to make it suitable for composting. Composting is a processed form of MSW management, where significant screening, sorting, and in some cases size reduction are performed to improve the handling characteristics and composition of the materials to be used for composting (Tongnetti et al. 2007). Recently, markets have been identified for MSW biowaste, such as landfill restoration, but whether the composting process is used as a pretreatment for disposal or in the production of a usable material, the output must be stable—that is, there are no further reductions in organic matter content—so that producers and users can be confident that contaminants in the material, such as heavy metals, have reached their maximum concentration. However, the definition of compost and biowaste stability is a matter of some debate. Belyaeva and Haynes et al. (2010) define compost stability in terms of microbial activity and the aerobic respiration rate as determined by the quantity of carbon dioxide produced by the microbes within the material. Similarly, other authors have related compost stability to microbial activity, but others relate stability to the loss of biodegradable material. The composting process, which is also used to decompose the organic fraction in MSW to produce biowaste, is the aerobic degradation of organic materials by a broad range of microorganisms including bacteria, actinomycetes, yeasts, and other fungi (Huang et al. 2010).

5.4.1.5 Monitoring and Parameter Adjustment

1. *Temperature*: Temperature is perhaps the most contentious of all the parameters controlling the rate of composting. Ultimately, when organic waste undergoes decomposition, the insulating effect of the material leads to the conservation of heat, and a marked rise in temperature takes place under normal composting conditions with optimum moisture content and aeration. The entire composting process may be divided into four stages: mesophilic, thermophilic, cooling, and maturation. Temperature monitoring in a composting mass may indicate the amount of biochemical activity taking place. However, in wheat straw and jamun leaf compost, only a low rise in temperature to a maximum of 45°C–50°C was observed (Genyanst and Lei 2003). As the decomposition is initiated, mesophilic microorganisms multiply and the temperature rises rapidly from its initial ambient level. Among the products of this initial decomposition are organic acids, which cause a drop in pH. As soon as the mesophilic range of temperature is exceeded, the activity of mesophiles drops and the process is taken over by thermophilic organisms. Volokita et al. (2000) investigated studies on the biological decomposition of wheat straw and reported a maximum temperature rise during the first 10 days of composting; after 60 days, there was a steady decline in temperature in all treatments excluding controls. Purohit et al. (2003) studied the composting of yard waste and water hyacinth in the tropics and found that the temperature inside the piles measured during the 16 days of the experiment was in the 40°C–60°C range, in which thermophilic bacteria and the cellulolytic fungus *Chaetomium thermophilum* were active in the decomposition process.

   The temperature later decreased to 30°C–40°C, and the mesophilic bacteria and the actinomycetes became predominant and the composting rate decreased accordingly. Moderately thermophilic

microorganisms have more thermal tolerance to superoptimal temperatures. Although an elevated temperature results in the destruction of
pathogens, it also reduces the level and activities of desirable microorganisms that are significant in the composting process; only in the last
few decades has the role of heat and elevated temperature been eluci-
dated (Finstein and Miller 1985). Belyaeva and Haynes (2010) reported
that the process of composting occurs in two stages: the first is the thermophilic stage, in which an increase in temperature occurs at around
65°C. At this stage, there is the decomposition of readily degradable
compounds such as sugars, fats, and proteins. During this stage, the
organic compounds degrade to carbon dioxide and ammonia with the
consumption of oxygen. Chen and Inbar (1993) reported that pathogenic microbes and helminth eggs are eliminated as a result of heat
generated during the process of composting. Thus, organic compost
is safer for use by farmers. The thermophilic stage is governed by the
fundamental principle of heat and mass transfer and by the biological constraints of living microorganisms. The rate of heat production
by organic material decomposition depends on the chemical, physical, and biological properties of the composting materials (Tiquia 2002,
Ghaly et al. 2006, Awasthi et al. 2014).

2. *pH*: pH is also an important factor determining the quality of compost. pH is a measure of the acidity or alkalinity of the mixture.
Decomposition will occur most readily in a neutral medium because
most microorganisms grow best under neutral conditions. Under
aerobic conditions, a material that is initially neutral will experience
a decrease in pH from the start of composting. This will typically
be followed by an increase in pH that will result in a final state that
is slightly alkaline. Most materials decomposing aerobically will
stay within a pH range that is conducive to microbial growth, thus
eliminating any need for pH control (Eriksen et al. 1999). There is
an intrinsic relationship between temperature, pH variation, and
time during composting (Ghosh 2004). The early mesophilic stage
shows a decrease in pH (acidic), and with an increase in the temperature of the composting mass, there is a corresponding increase in
pH. A maximum pH of around 8 synchronizes with a temperature
peak, with a subsequent leveling off at alkaline pH (Bhawalkar and
Bhawalkar 1993). In the initial stage, the pH value decreases due to
the production of organic acids derived from the intensive fermentation of carbohydrates. Afterward, the pH begins to rise and reaches
9 in the thermophilic period. This increase results from the release
of ammonia due to the start of the proteolytic process. Chefetz et al.
(1998) reported that the pH drops temporarily during the thermophilic stage due to the accumulation of organic acids, which highlights rapid microbial activity, and then these acids are used as a

substrate by other microorganisms. During the cooling down and maturation stages, the pH drops to a neutral value. pH cannot be considered a proper parameter to assess compost maturity as its overall trend is not describable by monotonic function (Chang et al. 2006).

3. *Moisture*: Improper moisture content can create problems for the degradation of organic substrates as well as the growth and multiplication of microbes. A dry pile is not detrimental to microbial activity, but it forms dust that carries odors and possibly fungal pathogens such as *Aspergillus fumigatus*. To achieve the optimal decomposition rate, the moisture content of the organic material should be between 50% and 60% by wet weight (Spellman and Whiting 2007). If the moisture content is in excess of 60%, the compost becomes soggy, thereby reducing the amount of air present, leading to anaerobic conditions. A moisture content of less than 45% is inadequate to satisfy the needs of microorganisms. Low moisture content also lowers the temperature of the entire compost mass. However, the moisture requirement has been found to range between 40% and 80%, depending on the nature of the organic material to be composted (Zurbrugg et al. 2004; Wong et al. 2009). Several researchers have reported that maintaining 50%–60% moisture content is ideal for composting organic MSWs (Garcia-Gil et al. 2000).

4. *Oxygen*: Composting systems are distinguished by oxygen usage; that is, either they are aerobic or anaerobic in nature. Aerobic decomposition occurs at a rapid pace and progresses at higher temperatures; in addition to this, aerobic conditions do not produce foul odors, while anaerobic decomposition is conducted in a sealed environment to remove foul odors. Compost piles are mixed at intervals for their proper aeration, but sometimes it becomes difficult to determine the exact periods for the aeration process. Excessive aeration is not harmful to the compost mix, except that an ideal temperature range is harder to maintain, and excessive evaporation of the mix may cause a low moisture content (Ahmad et al. 2007). The optimum oxygen level required for operation is around 10%–30% (White et al. 1997). Thus, aeration is a major consideration in most composting techniques due to the advantage gained by maintaining aerobic conditions. Under farmhouse composting conditions, this can be accomplished by the periodical turning or mixture of the composting mass (Warman and Termeer 2005).

5. *Odor generation*: Odor generation is one of the major concerns for any composting facility. It is better to understand the different ways in which odors can be formed so that ecological conditions can be manipulated for their prevention and treatment. Odors produced at the beginning of the composting period are generally caused by

the nature of the material used. Feedstocks such as manure or fish-processing wastes produce strong odors at the beginning of composting that diminish as composting proceeds. Various odorous compounds are released through either biological (microbial respiration) or nonbiological processes (chemical reactions). Odors can be in gaseous form, or these gas phase odorous compounds can also be adsorbed by particulates such as dust. Volatile organic compounds (VOCs) are generated during the decomposition of long-chain carbon compounds and are also found as intermediates in carbohydrate metabolism. Some of the common odorous VOCs are volatile fatty acids (VFAs), indoles, phenols, aldehydes, amides, amines, esters, ethers, and ketones (He et al. 2001). Odorous sulfur compounds include gases such as hydrogen sulfide (H_2S), dimethyl sulfide (CH_3 SCH_3), methanethiol (CH_3SH), ethanethiol (CH_3CH_2SH), propane-thiol ($CH_3CH_2CH_2SH$), mercaptans, and other compounds that are formed from the decomposition of sulfur-containing compounds in anaerobic conditions (Graves et al. 2010). Sulfur compounds are also produced nonbiologically through the reaction of various compounds that accumulate in the compost pile.

For the removal of odor, biofilters can be the most efficient technology. They use microorganisms to decompose odorous organic compounds. Most of the odor-generating compounds are metabolic intermediates that are further metabolized and utilized by microorganisms (Graves et al. 2010). Soil filters can also be used to control odor. The soil in the filter removes odor through the chemical absorption, oxidation, filtration, and aerobic biodegradation of organic gases. Filters require fine-textured soil, sufficient moisture, and a pH range of 7.0–8.5 for proper functioning (Graves et al. 2010).

6. *Odor management*: Odor is the most efficient and simple indicator of whether the pile conditions are aerobic or anaerobic. Different types of odors replace the unusual odor of the substrate within a few days after the initiation of the process. If the substrate is MSW, the odor comes from raw garbage. However, if conditions are unsatisfactory (e.g., anaerobiosis), the overpowering odor would be that of putrefaction. If the C:N ratio of the substrate is lower than about 20:1 and the pH values are above 7.5, the odor of ammonia could become predominant (UNEP 2001). At low temperatures, strong, putrid odors that sometimes smell of sulfur indicate anaerobic activity.

If ammonia odors are produced by the compost pile, then it may be necessary to conserve nitrogen to stop nutrient loss from the compost. This type of management technique includes the addition of carbon-rich material to compost and a reduction in the turning frequency (Graves et al. 2010).

5.4.1.6 Main Types of Composting System

Compost systems currently in vogue can be classified into three broad categories:

- Windrows
- In-vessel systems
- Vermicomposting

1. *Windrows*: Windrows are a simple composting technology requiring minimal engineering design or labor and are a widely used method for the composting of yard trimmings and MSW. They require minimum investment regarding equipment and finance. Windrow composting is an inappropriate technology for MSW because of odor emission and the attraction of vectors (e.g., rodents, flies, etc.). Hence, it is progressively being abandoned but increasingly being used for garden waste only. As one would suspect, the designation *windrow systems* reflects the distinguishing feature of such systems—namely, the use of windrows. There are two types of windrow system: *static* (stationary) and *turned*. The principal difference between the static version and the turned version is the fact that in the static version, aeration is accomplished without disturbing the windrow, whereas in the turned version, aeration involves demolishing and reconstructing the windrow (UNEP 2005). The principle steps involved in the windrow composting process are
   - Inclusion of a bulking agent into the waste
   - Construction of the windrow
   - Composting
   - Screening of the composted mixture
   - Curing and storage

   For proper aerobic conditions and uniform decomposition of waste, compost piles are frequently turned so that the cooler outer layer is exposed to higher temperatures and the aerobic composting process takes place uniformly. Turned windrow operation is conducted under shelter or in an outside area on a firm surface to stop the percolation of leachate into the soil (Tiquia et al. 1996; Wong et al. 2001). The size of the windrow reduces with time with the decomposition of material into a simpler form. Windrow height depends on the feedstock, the season in which composting is conducted, the region, the compaction capacity of the compost material, and the turning equipment. To prevent excessive heat and to insulate the composting material, an ideal windrow height of 1.5 to 1.8 m is suggested.

2. *In-vessel systems*: In-vessel composting refers to a group of methods that confine the composting materials within a building, container, or vessel. In-vessel systems attempt to create optimum conditions for the microorganisms, thereby giving improved control of the composting process and accelerating decomposition. As in all composting systems, the supply of air to all the material being composted is the primary factor that determines the effectiveness of the process.

In-vessel systems can be used for the treatment and handling of a large amount of organic waste (mainly MSW scale), and the resulting stabilized form can be used as soil amendment. This system can also be used with aerated static piles, with the inclusion of removable covers over them. To minimize and control foul odors, a higher C:N ratio is maintained, with proper ventilation facilities to increase aeration. Coarse-grade carbon material is also used to allow better air circulation through the compost. Biofilters are provided to prevent and capture any naturally occurring gas from aerobic composting. Advanced systems are designed to reduce odor issues, and the integration of an in-vessel system with anaerobic digestion will result in higher energy and resource output. In this type of system, organic matter is first passed through anaerobic digestion and then composting is done under forced aerated conditions. In-vessel systems take 14 days to produce stabilized compost (UNEP 2005).

Various types of in-vessel systems have been developed over the years, such as the Dano drum, the naturizer system, the metro and channel types, and the Fairfield reactor. The primary aim of designing an in-vessel reactor is to accelerate the composting process and to produce an ambient environment by the elimination of undesirable conditions (UNEP 2005). The aeration design of an in-vessel system consists of features such as forced aeration, stirring, and tumbling. Forced aeration is conducted in most of the systems, whereas stirring is done via the rotation of plows or augers. Tumbling can be done by dropping compost material from one level to a lower level (UNEP 2005).

3. *Vermicomposting*: Vermicomposting is a process that uses red earthworms to consume organic waste, producing castings (an odor-free compost product that can be used as mulch), soil conditioners, and topsoil additive. Bacteria and millipedes help in the aerobic degradation of the organic material. Vermicomposting is especially useful for processing food waste since the worms consume the material quickly and there are fewer problems with odor. Vermicomposting does not generate temperatures high enough to kill pathogens. For this reason, vermicomposting is more appropriate for food, paper, and yard waste. The most common types of earthworms used for vermicomposting are brandling worms (*Eisenia fetida*) and red

worms or red wigglers (*Lumbricus rubella*) (Yadav and Garg 2011; Khwairakpam and Bhargava 2009). These worms are generally found in aged manure piles and have alternating red and buff-colored strips. In this process, various worms and microorganisms are involved at different stages of compost development; this also contains various worm castings. Earthworm castings in the home garden often contain 5–11 times more nitrogen, phosphorous, and potassium than the surrounding soil (Baoyi et al. 2013; Sen and Chandra 2006). Red worms in vermicompost act in a similar fashion, breaking down food wastes and other organic residues into nutrient-rich compost. The nutrients in vermicompost are often much higher than traditional garden compost. Finished vermicompost should have a rich, earthy smell if properly processed by worms. Vermicompost can be used in potting soil mixes for house plants and as a top dressing for lawns. Screened vermicompost combined with potting soil mixes makes an excellent medium for starting young seedlings. Vermicompost also makes an excellent mulch and soil conditioner (Li et al. 2011). Parameters that can be considered for the selection of appropriate vermicomposting technology include

- The amount of feedstock for processing
- Availability of funds
- Site and space restrictions
- Climate and weather conditions
- Regulatory restrictions
- Facilities and equipment on hand
- Availability of low-cost labor

Vermicomposting is of four types.

- Windrows
- Wedge systems
- Bed and bin systems
- Reactor systems

Windrows are employed both open and under cover, but the major disadvantage is that they require a large surface area. It is difficult to conduct the vermicompost process without earthworms, so a mechanical harvester is commonly used for these operations. The wedge system is a modified windrow system that maximizes space and simplifies harvesting, as there is no need to separate the worms from the vermicompost. Organic materials are placed at an angle of 45° against the finished windrow. The piles can be placed inside a covered space or outdoors if it is covered with a tarp to avoid the leaching of nutrients. A front-end loader can be used to build a windrow 1–3 m wide by whatever length is appropriate (Yadav and Garg 2009; Xing et al. 2012). The windrow is started by spreading a 30–45 cm layer of organic materials the length of one end of the available space. Up to 500 g of red worms are added per square meter of windrow surface area. Subsequent layers of 5.0–7.5 cm of organic material are added weekly, although 7–15 cm layers can be added in colder weather. After the windrow reaches 0.6–1.0 m thick, it can be led sideways by adding the next layers at an angle against the first windrow (Shermann-Huntoon 2000). Once this is achieved, worms start migrating from the first windrow to the fresh windrow. It is a labor-intensive process to harvest worms and vermicompost by hand. The major advantage associated with hand-driven vermicompost is that when the worm beds get too hot, the worms can burrow deeper to where the temperature remains below 23°C, and then the system is kept undisturbed for 3 days, compared with automated reactors, which require daily inspection for moisture and temperature levels (Shermann-Huntoon 2000). The disadvantage is that the worms and castings must be separated manually. Feed stocks are added daily in layers on top of the mesh or grate. Finished vermicompost is harvested by scraping a thin layer from just above the grate, which falls into a chamber below. These systems can be relatively simple and manually operated or fully automated with temperature and moisture controls. For maximum efficiency, they should be under cover.

The advantages of vermicomposting are as follows.

- Earthworms efficiently break and fragment the organic waste into a stable, nontoxic material with high economic value as a fertilizer.
- Properly produced vermicompost has excellent structure, porosity, aeration, drainage, and moisture-holding capacity.
- Vermicompost supplies a suitable mineral balance, improves nutrient availability, and can act as complex fertilizer granules.
- As with the composting process, vermicomposting provides a significant reduction in waste bulk density, although this tends to take longer.
- The low-technology systems can be easily adapted and managed on small farms or livestock operations.
- Vermicomposting produces compost in about 3 weeks.

The main disadvantages are that vermicomposting requires more management and maintenance than other composting systems to maintain healthy worms, and it has not yet been successfully used for large-scale treatment. The environmental effects are similar to composting, and windrows may need proper designing to prevent leachates. The odor and leachate problems are controlled in other types of vermicomposting.

5.4.2 Biomethanation

Biomethanation is a process in which the organic matter of solid waste is decomposed by microorganisms to produce biogas under anaerobic conditions (i.e., without air). Biogas includes various gases: mainly methane, carbon dioxide, hydrogen, and hydrogen disulfide. In this process, oxygen is not required for the decomposition of waste; the anaerobic process is inherently the most energy-efficient option for the safe disposal of garbage. All other options are either energy intensive, inconvenient, or environmentally unsafe. A wide variety of process applications for the biomethanation of wastewater, slurries, food waste, and solid waste have been developed (Angelidaki et al. 2011). For over 100 years, this technology has been used in different practical situations, demonstrating itself as a viable platform. Nevertheless, technical improvements are in progress (EPA 2008). The characteristic properties of biogas are mainly affected by changes in pressure and temperature; moisture content also plays a significant role in biogas production. The following is an explanation of the anaerobic process:

- Anaerobic degradation of organic material is a two-stage process in which complex organic material breaks down into short-chain volatile fatty acids, which is then further converted to methane and carbon dioxide.
- The complex polymer and acids are fermented at the same time in a single-phase system, whereas the separation of acidogenic and methanogenic bacteria occurs in a two-phase system. The retention period of material in a biomethanation chamber depends on the chemical characteristics and the design of the system (single stage, two stage, multistage). The generalized formula of the anaerobic process is:

$$\text{Organic matter} + \text{anaerobic bacteria}$$
$$\rightarrow CH_4 + CO_2 + NH_3 + H_2S + \text{other end products}$$

- A highly temperature-controlled process takes less time in the treatment of waste. Mainly, this process is suitable for wet, semiliquid waste such as food or sewage sludge, the composting of which is difficult because of improper air circulation (CED 2012).

As this process is mildly exothermic in nature, it requires extra heating during thermophilic conversion; the higher the temperature, the shorter the retention time (Figure 5.4).

The anaerobic digestion process is of two types.

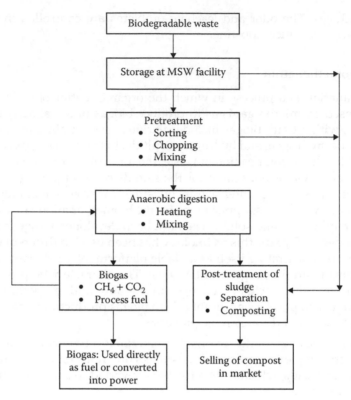

FIGURE 5.4
The anaerobic digestion process for biodegradable waste. (From William, *Waste Treatment and Disposal*, pp. 325–342. New York: Wiley, 2005.)

1. *Dry anaerobic process*: This process is adopted when the total solid content in waste is above 25%; here waste is degraded in a single stage, either in batch or continuous mode. In this process, much less water is required or heated to maintain thermophilic conditions (500°C–600°C) for better growth of bacteria. This process takes around 15–20 days and is followed by postdigestion for residue stabilization and maturation (CED 2012).

2. *Wet anaerobic process*: This process is adopted when the total solid content is 3%–8% (CED 2012). This is a completely mixed mesophilic reaction. As it is a single-stage process, a lot of problems occur such as the formation of scum layers and there is difficulty in keeping the content thoroughly mixed. Another problem that occurs in this process is the presence of two different types of microorganisms—that is, acidogenic and methanogens both require different pH values of medium for proper functioning. This problem can be solved by running the process in a two-stage digester as hydrolysis and acidification.

5.4.2.1 Different Biochemical Processes of Biomethanation

Methane and carbon dioxide are the primary products of the process, and are generally produced by the decomposition of organic matter. Biogas formation is a four-step process in which material is broken down from complex to simpler forms in each step in the presence of different microorganisms. The four steps are as follows: hydrolysis, acetogenesis, acetogenesis, and methanogenesis (Figure 5.5).

All the processes shown in Figure 5.5 take place simultaneously in the tank. The speed of the whole biomethanation process depends on the speed of the slowest reaction in the tank.

1. *Hydrolysis*: This is the first step of the biomethanation process in which matter such as polymers break down to monomers and oligomers. In this process, polymers such as carbohydrates, lipids, and proteins convert into glucose, glycerol, purines, and pyridines. Microorganisms in this process release hydrolytic enzymes, converting biopolymers into simpler and more soluble compounds.

$$Lipids \rightarrow fatty\ acids \rightarrow glycerol$$

$$Proteins \rightarrow amino\ acids$$

   The compounds produced during the hydrolysis process are further decomposed by microorganisms involved in other steps and are used for their metabolic processes (Teodorita et al. 2008).

2. *Acidogenesis*: In this process, a leftover of the hydrolysis process is further broken down by fermentative bacteria into methanogenic products. Pure sugar, amino acids, and fatty acids are mainly converted into acetate, carbon dioxide, and hydrogen with a portion of volatile fatty acids and alcohol. The essential acids produced are acetic acid, butyric acid, propionic acid, and ethanol (Monnet 2003).

3. *Acetogenesis*: Methanogenic bacteria cannot decompose volatile fatty acids and alcohol with long carbon chains; in this step, they

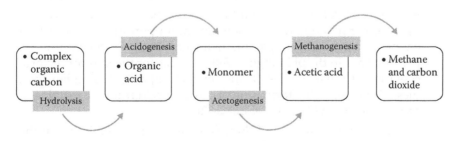

FIGURE 5.5
The main processing step of anaerobic digestion.

are oxidized into methanogenic substrates such as acetate, carbon dioxide, and hydrogen. Both acetogenesis and methanogenesis run simultaneously as a mutualism of two groups of microorganisms.

4. *Methanogenesis*: This is the most vital step of the whole anaerobic digestion process, as the rate of methanogenesis reaction is slowest among all other reactions. Methane can be generated in two ways, either by using the cleavage of acetic acid molecules to produce carbon dioxide and hydrogen or by the reduction of carbon dioxide with hydrogen. This process is affected by many factors, such as a change in temperature or pH, excessive loading in a digester, and varying components of feedstock (Teodorita et al. 2008).

Acetic acid → methane → carbon dioxide

Hydrogen + carbon dioxide → methane + water

5.4.2.2 Parameters Affecting Anaerobic Digestion

Certain parameters affect the various reactions of anaerobic digestion. All the parameters need to be controlled accurately to obtain an optimum yield of biogas.

1. *Temperature*: The biomethanation process is mainly affected by temperature variations. The process happens in three different temperature ranges: psychrophilic (below 25°C), mesophilic (25°C–45°C), and thermophilic (45°C–70°C). The temperature of the digester decides the retention period of feedstock: the higher the temperature, the faster the microbial activity, and the shorter the retention period. Most of the modern biogas plants work in the thermophilic temperature range due to the many advantages of the thermophilic process, such as the higher growth rate of methanogenic bacteria, better degradation of solid substrate, improved digestion, and so on. The thermophilic temperature range also has some disadvantages, such as energy demand increases due to increases in temperature; also, there is a high risk of ammonia inhibition (Teodorita et al. 2008).

2. *pH value*: pH value plays a vital role in the biomethanation process, as various microorganisms show optimum performance in different pH ranges. For example, methanogenic bacteria show optimum performance in the pH range of 7–8, whereas for the mesophilic process the pH range is 6.5–8.0, and for thermophilic it is higher than mesophilic due to the formation of carbonic acid from the reaction of carbon dioxide with water (CED 2012).

3. *Nutrient concentration*: The C:N ratio is the relationship between the amount of carbon and nitrogen present in any compound for its growth. The optimum C:N ratio for the biomethanation process is

between 20:1 and 30:1, and the ideal C:N ratio is 25:1 (Ramasamy and Nagamani 2010). Biogas production is highly affected by changes in the C:N ratio; higher values show lower gas production, whereas very low values indicate ammonia accumulation, which in turn raises the pH value above 8.5. As a result, an unsuitable environment is created for methanogenic bacteria.

4. *Organic loading rate*: The biological conversion capacity of the anaerobic digestion process is defined by the organic loading rate. If the feedstock is added above the acceptation rate of the digester, then it yields low gas volumes and accumulates high amounts of unwanted substances, such as fatty acids. Monnet (2003) expressed the organic loading rate in kilogram chemical oxygen demand or volatile solid (VS) per cubic meter of the reactor.

5.4.2.3 Types of Anaerobic Digester

Depending on the organic loading rate and the amount of biogas required, a variety of digesters have been designed. In different kinds of digester, organic waste is loaded either continuously or intermittently.

1. *Standard-rate single-stage digester*: In this type of digester, untreated organic matter is added to the digestion zone for degradation. Here, external heat is provided to raise the temperature of the digester for gas production; as the gas rises it brings up slurry particles and various minerals such as grease and fats to the top, forming a thick scum layer. Due to the stratification of the scum layer at the top, it becomes difficult to mix the slurry; for this reason, single-stage digesters are used for small-scale applications. The detention time for this process varies from 30 to 60 days (CED 2012).

2. *Multistage digester*: In this digester, the anaerobic process occurs in different reactors so that various reactions occur in a more controlled and flexible manner. Two reactors are used: one for hydrolysis/acidogenesis and another for methanogenesis. The primary benefit of having different reactors is to provide a certain degree of control over the rate of hydrolysis and methanogenesis. This system provides better biological stability for the rapid degradation of organic matter (Monnet 2003).

5.4.3 Thermal Processing of Municipal Solid Waste

Thermal processing is burning out a combustible portion of solid waste in the presence of air or oxygen, to recover energy from solid waste as heat and steam. There are various processes for the thermal treatment of solid waste, such as combustion, gasification, pyrolysis, RDF, and so on (Figure 5.6).

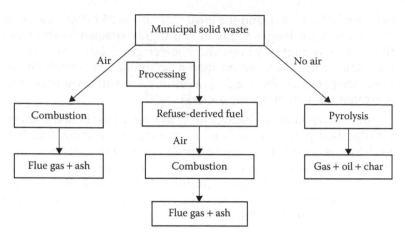

FIGURE 5.6
MSW thermal processing options.

5.4.3.1 Incineration

Incineration is a process in which the chemical elements of the solid waste mix with oxygen or air in the combustion zone to generate heat. This process occurs in the presence of excess air for the proper mixing and combustion of waste. The excess air required for incineration depends on the moisture content, the heating value, and the type of technology used for the combustion of waste. In addition to oxygen, carbon and hydrogen are also necessary for the combustion process; they are either formed in their free or combined states. However, the presence of sulfur is a matter of concern as it combines with oxygen to form sulfur dioxide, which pollutes the surrounding environment (CED 2012). The generalized formula for the combustion reaction is as follows:

$$\text{Solid waste} + \text{air} \rightarrow CO_2 + H_2O + O_2 + NO_x + \text{energy}$$

Various intermediate reactions occur during the whole combustion process. These intermediate reactions occur due to the high temperature of the combustion chamber. The different reactions that happen during incineration are listed in Table 5.3.

Incinerators operate in the range of 900°C–1100°C. This temperature range ensures complete combustion, eliminates odor, and protects the reactor's wall (CED 2012). To increase the overall efficiency of the incineration process—that is, heat output and gas emission—it is necessary to balance excess air with temperature, time, and turbulence. Incineration technology is designed to reduce the overall volume of solid waste to 10%, which is left at the bottom of the combustion chamber in the form of ash, which is inert in nature and usually landfilled. Compared with conventional fossil fuel–based technology, the waste-to-energy (WTE) combustion process saves significant potential

TABLE 5.3

Chemical Reactions of the Combustion Process

Combustible Elements	Reaction
Carbon	$C + O_2 \rightarrow CO_2$
Hydrogen	$2H_2 + O_2 \rightarrow 2H_2O$
Sulfur	$S + O_2 \rightarrow SO_2$
Carbon monoxide	$2CO + O_2 \rightarrow 2CO_2$
Nitrogen	$N_2 + O_2 \rightarrow 2NO$
Nitrogen	$N_2 + 2O_2 \rightarrow 2NO_2$
Nitrogen	$N_2 + 3O_2 \rightarrow 2NO_3$
Chlorine	$4Cl + 2H_2O \rightarrow 4HCl + O_2$

Source: Hecklinger, *The Engineering Handbook*, Boca Raton, FL: CRC Press, 1996; Velzy and Grillo, *Handbook of Energy Efficiency and Renewable*, Boca Raton, FL: CRC Press, 2007.

resources; for example, 1 ton of MSW is equivalent to 2.5 tons of steam at 400°C and 40 bar pressure. Similarly, 1 ton of MSW holds energy equivalent to 200 kg of oil, and 500 kWh electricity can be produced from it (Jayaram 2011).

Despite the fact that combustion technology produces high amounts of energy and reduces the waste volume to a large extent, the technology is not being utilized in a number of nations around the world because of the high risk of environmental pollution due to the emission of toxic gases (Jayaram 2011). In addition to this, there are other concerns associated with incineration systems, such as leachate, scrubber effluents, and the disposal of ash that contains heavy metals. The reason for the failure of most WTE plants in developing countries is the low calorific value of MSW, as incineration systems are designed for high-calorific-value waste, which includes a large proportion of paper, plastics, and so on. Also, the poor quality of solid waste causes the emission of fine particulate matter, oxides of nitrogen and sulfur, heavy metals, carbon monoxide, and so on. To remove these toxic products of the combustion process, modern WTE facilities are installed with equipment such as flue gas cleaners consisting of scrubbers, electrostatic precipitators, and so on.

The primary aim of any incineration process is to reduce large volumes of waste, recover the maximum energy from solid waste, stabilize waste, and remove toxicity from waste. A large number of projects have failed in past decades in developing nations because of the poor quality of waste. Therefore, it is necessary to understand the various planning issues related to incineration project development (Figure 5.7).

5.4.3.1.1 Types of Incineration System

There are three types of incineration system: mass-burn combustion, modular incineration, and fluidized-bed incineration. Of these, fluidized-bed systems are not much used throughout the world, whereas RDF-based systems have been used in developing nations; various case studies have found that

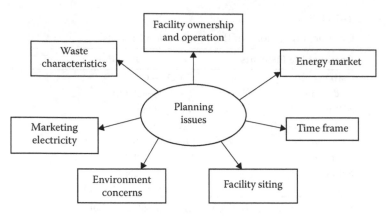

FIGURE 5.7
Planning issues related to incineration project development.

RDF-based incineration systems shut down in most developing countries because of mixed solid waste.

1. *Mass-burn combustion*: This type of system has a reciprocating grate system for the combustion of waste, a refractory lining, and a water-walled steam generator. It contains two to three incineration units with a capacity of 50–1000 MT solid waste per day. Before the combustion of solid waste, it undergoes preprocessing, where oversized materials are removed without any extraction of recyclables. The primary benefit of mass-burn combustion is that it generates a higher quality of steam, which is then utilized to generate electricity. Otherwise, an extraction turbine can be used to produce electricity as well as processed steam, which can be used for other heating purposes.

2. *Modular incineration*: This system is made up of prefabricated units with a relatively small capacity of 5–120 MT solid waste per day. This system typically contains one to four units, which makes a total capacity of 15–400 MT per day (CED 2012). The combustion of solid waste is different from the mass-burn process; the system contains two combustion chambers, and the process is performed in two stages. The primary benefit of this system is due to the prefabrication units, which make the construction of the facility simple and save time. This system is used for small-scale industries and communities. However, this type of system is becoming unpopular because of inconsistencies in operation and inadequate air pollution control.

3. *Fluidized-bed incineration*: This kind of incinerator has a bed of solid particles made up of limestone or sand to withstand high temperatures and is connected with an air distribution system where gas flows through the bed with varying velocity, causing it to bubble. The

two types of bed are bubble beds and circulating beds; both work on different air velocities and have different bed materials. Also, both systems have implications for the types of waste combusted as well as heat transfer to the energy recovery system (Jayaram 2011). In comparison with other incineration technologies, fluidized-bed incinerators can burn solid waste with different moisture and heat contents. This system is more efficient than other technologies as it is more energy efficient, more consistent in operation, generates fewer residues, and has lower air emissions. Due to its high compatibility with recycling systems, this technology can be implemented at small scales in cities with high-recycling facilities in developing nations (Jayaram 2011).

5.4.3.2 Refuse-Derived Fuel

RDF is generated from the mechanical processing of mixed MSW in which the noncombustible portion of solid waste is removed to produce the most homogeneous mixture. In comparison with mass-burn facilities, RDF-based facilities recover valuable recyclable products from mixed waste before converting them into pellets or fluff. RDF systems primarily perform two functions: production and combustion; production facilities remove recyclables such as glass and ferrous materials and reduce the size to produce different types of RDF such as fluff, pellets, and bricks. The most suitable waste for RDF production should have a high carbon content after the separation of recyclables (Figure 5.8).

5.4.3.2.1 Types of RDF System

1. *Shred and burn system*: This is the simplest RDF system, in which minimal processing of unprocessed solid waste is done, such as the removal of ferrous materials. In this system, there is no provision for the removal of the noncombustible portion of solid waste. After

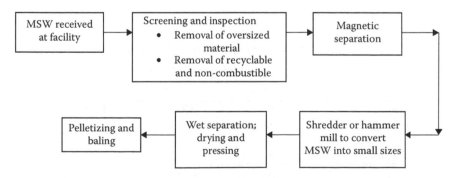

FIGURE 5.8
Process showing the production of RDF at an MSW facility.

minimal processing, solid waste is shredded to the required particle
size and then sent to the combustor.

2. *Simplified process system*: In this type of RDF system, mixed solid waste
is processed mechanically to remove noncombustible, recyclable, and
ferrous materials. After proper processing, solid waste is finally put
into the shredder to create homogeneous particles 10–15 cm in size for
optimal energy recovery during the combustion process (CED 2012).

In developing nations, MSW is found in heterogeneous forms with low
heat value and high moisture content, due to which mass-burn facilities
cause air pollution, whereas RDF-based incinerators solve this problem to
some extent. Despite the fact that RDF facilities remove contaminants and
recyclables from the combustion process, the production of RDF is a complex
process that increases the operation and maintenance cost of the plant and,
hence, reduces the reliability of the facility.

5.4.3.3 Pyrolysis

The pyrolysis process is also called *destructive distillation*. Some solid waste management professionals consider gasification and pyrolysis the same process. In
reality, they have significant differences: pyrolysis uses an external source of
heat to perform the endothermic pyrolytic reaction in an oxygen-free environment, whereas gasification converts waste into gaseous substances using air
but shows a partial combustion reaction. The pyrolysis process is highly endothermic in nature, whereby organic, thermally unstable substances break down
through thermal cracking and condensation reactions in gaseous, liquid, and
solid substances. The most suitable waste for pyrolysis contains plastics, paper,
and biomasses composed of the major polymeric chains of cellulose, hemicellulose, and lignin. On thermal degradation, these longer, complex chains convert
into simpler stable molecules, thus resulting in the formation of oil that can be
used as fuel (William 2005). The pyrolysis process is classified into three different classes based on operating parameters (Table 5.4).

TABLE 5.4

Types of Pyrolysis Process Based on Different Parameters

Parameters	Types of Pyrolysis Process		
	Slow	Fast	Thermolysis
Temperature (K)	550–900	850–1250	1050–1300
Heating rate (K/s)	0.1–10	10–200	>1000
Particle size (mm)	5–50	<1	<0.2
Retention time (s)	300–3600	0.5–10	<0.5

Source: Agarwal, M. 2014. An investigation on the pyrolysis of municipal
solid waste. In PhD thesis from the School of Applied Sciences,
College of Science, Engineering and Health, RMIT University.

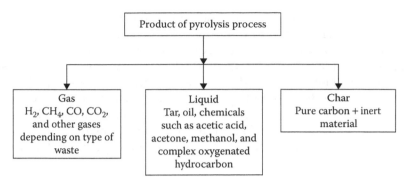

FIGURE 5.9
Products of the pyrolysis process.

Pyrolysis products can be utilized for various purposes; for example, oil can be utilized directly in fuel applications or after being upgraded to refined oil, and char can be used as carbon black, activated carbon, and char oil, or as char water slurry for fuel. The products of the pyrolysis process can also be used as chemical feedstock (Figure 5.9).

5.5 Reuse and Recycling

To reduce a large amount of mixed waste flow into open dump sites and landfill, it is necessary to incorporate the 3Rs—reduce, reuse, and recycle—in the solid waste management system of any city. Reducing, recycling, and reusing materials and packaging at home, work, and public places will save usable material from entering into landfill. Reducing waste by purchasing products with little or no packaging or simply reducing the consumption of goods such as paper eliminates or reduces the amount of waste that enters the waste stream. This can reduce product costs and fossil fuel consumption and save natural resources. Reusing products such as durable goods through facilities such as swap shops or second-hand stores likewise accomplishes the same savings. Recycling reduces the cost and pollution associated with raw-material extraction and manufacturing dependent on large quantities of increasingly expensive carbon-emitting fossil fuels that contribute to climate change. Recycling also places an economic value on materials that degrade the environment through their toxicity, such as the lead in automotive batteries or simply the space they require in a landfill. Additionally, it helps preserve valuable natural resources. Recycling promotes the development of innovative products made from recycled materials, such as countertops and bathroom tiles made from recycled glass. Recycled plastic lumber used for decking and landscaping is made from recycled materials such as plastic milk jugs and is impervious to insects, moisture, and bacteria, outlasts treated lumber, and can be recycled back at the end of its useful life (American Chemistry Council 2015). Yard trimmings and unused food waste account for 26% of the solid waste that goes into landfills, and most of this waste can be composted and used to enhance soil in vegetable and flower gardens. Composting reduces the need for petroleum-dependent fertilizers, increases crop yields, and helps reduce plant diseases and pests, which in turn lessens the need for toxic pesticides, herbicides, and fungicides. Composting is simply a natural process in which microorganisms such as bacteria and fungi break down organic waste into nutrient-rich soil. Providing the right conditions simply speeds up this process.

5.6 Ultimate Disposal Methods

Landfill is the primary approach for the dumping of waste generated by commercial and industrial installations and households. Nowadays, waste management and recycling are considered suitable options in disposal methods, even though landfill is likely to continue. There are some other nonlandfill disposal approaches: inceneration, RDF manufacture, and composting play significant roles in the disposal of waste. These methods are considered for short-term objectives, but in the long run, landfill is conceived as being sustainable. Therefore, it is important that landfills should be designed, located, monitored, and operated to check that they are not imposing any significant threat to the natural environment and human health (Figure 5.10).

5.6.1 Types of Landfill

There are three basic types of landfill: conventional landfills for mixed MSW, landfills for shredded waste, and monofills for specialized waste (Tchobanoglous et al. 1993).

1. *Landfills for mixed waste*: Most of the landfills constructed in developing and developed countries are for mixed waste. In developing countries where source segregation is not practiced, high amounts of organic waste and recyclables reach landfills, causing air, soil, and groundwater pollution. This type of landfill also receives nonhazardous waste from industries and dry sludge from wastewater treatment plants (Tchobanoglous et al. 1993). For covering waste at landfills, native soil with the lowest permeability is used to stop water percolating through it. Sometimes, it is difficult to find soil of low permeability and the required compaction in the local area; in that case, demolition waste, compost produced from MSW, or old rugs can also be used for landfill cover.

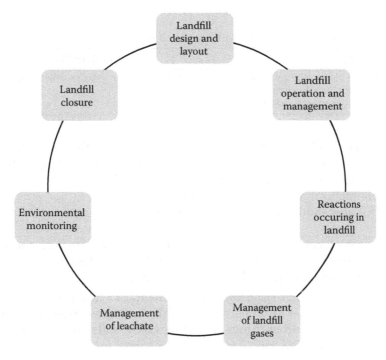

FIGURE 5.10
Principle elements of landfill planning, design, and operation.

2. *Landfills for milled waste*: This method is adopted in developed countries such as the United States. In this method, first the waste is shredded or milled to reduce the size of the solid waste before sending it to landfill. The main benefit of this process is that shredded waste can be compacted to 35% greater density than normal MSW; this results in the saving of more landfill area in regions where land cost is very high (Tchobanoglous et al. 1993).

3. *Monofills for specialized waste*: Monofills are generally constructed for individual constituents of waste that cannot be placed in mixed-waste landfills. This type of waste can be electronic waste or ash from incineration plants. Incineration ash monofills produce a sulfate odor that needs to be controlled by a gas recovery system (Tchobanoglous et al. 1993).

5.6.2 Environmental Impacts of Landfilling and Their Control

Landfilling is acceptable if properly conducted. If not carried out according to standard protocols, it can have negative effects on the habitat, categorized as short-term and long-term impacts (Table 5.5). The general objective

TABLE 5.5

Impact of Improper Municipal Solid Waste Landfilling

Short-Term Impacts	Long-Term Impacts
Noise, flies, odor, air pollution, unsightliness, and windblown litter. Such nuisances are generally associated with waste disposal.	Pollution of the water regime and landfill gas generation.

of environmentally acceptable landfilling is to reduce the short- and long-term influence or the mineralization of the deposited organic waste. Specific objectives include the prevention of surface water and groundwater pollution in the surrounding area of the landfill.

Landfill sites should only be selected after the following considerations:

1. *Economic considerations*: Economic considerations include the distance from waste generation to the landfill site, the size of the site, and the accessibility and availability of land.

2. *Environmental considerations*: Environmental factors include threats to the physical environment, particularly water resources. This comprises, among other things, site topography, soils, drainage, geohydrology, and land use.

3. *Public acceptance*: Public recognition is the primary factor in establishing a landfill in an area. It includes the potential negative influence of a landfill on human health aspects of life, and local land and property values. Local public persistence toward the establishment of a landfill can hinder its development. Meanwhile, economic and public acceptance considerations have proved to identify the normal area in which a landfill is cited. Within these constrains, the most favorable substantial environment option must be found.

5.6.3 Sanitary Landfilling with Biogas Recovery

5.6.3.1 Methods of Sanitary Landfilling

Sanitary landfilling is the practice of open crude dumping of MSW, which is appropriate in developing countries to manage the disposal of huge quantities of wastes because of the relative simplicity and flexibility of the waste management technology. It avoids hazardous impacts on human and environmental health of solid waste placed on land (ENVIS Center on MSWM 2016). Sanitary landfilling is an accomplished way to significantly reduce contact between the environment and waste, as waste is concentrated in a well-defined area. The result is good control of landfill leachate and limited access of vectors to the waste and gas. In modern waste management practices, recycling of organic waste and sustainable waste reduction processes, as well as strategies of landfilling, should be embraced.

Now a days, the implementation and practice of sanitary landfilling are seriously inhibited in most developing countries due to the inadequacy of reliable information about local contexts, as well as by insufficient funds and well-trained staff (Maier, 1998; Mane and Praveen 2013). There are three methods of sanitary landfill, that is, trench, area or ramp, and valley and ravine area.

1. *Trench method*: In this method, sanitary landfills are constructed over plain and moderately sloping ground. Solid waste is placed and well compacted in shallow excavated trenches around 6.5–7.5 m wide and twice as wide as the size of any compacting equipments (Salvato 1992). The depth of the trench is calculated from the depth of groundwater and the establishment of finished grade.

2. *Area or ramp method*: This method is adopted on rolling terrain and flat surfaces. The width and slope of landfill depends on three factors: the nature of the terrain, the amount of waste received per day, and the number of trucks unloaded in a particular time period. To achieve the proper compaction of waste and avoid the scattering of debris, the working face should be made small. In this method, cover materials are taken from nearby areas or other sources (Salvato 1992).

3. *Valley and ravine area method*: In these two methods, the ravine method is normally the best operational method. In those areas where a ravine is very deep, the solid waste should be placed in lifts from the bottom up to a depth of 8–10 ft (Salvato, 1992), while cover material is taken from the sides of the ravine. In addition, it is not always necessary to extend the lift from the entire length of the ravine, which may be urgently needed to construct the first layer for a relatively short distance from the head of the ravine across its width. Meanwhile, the length of this initial lift should be determined so that there is 1 year of settlement before the next lift is properly placed (Hilary et al. 2002), however this is not essential if the operation can be well controlled. The succeeding lifts are constructed by trucking solid waste over the first lift to the head of the ravine.

5.6.3.2 Leachate Collection System

Leachate collection and removal are essential requirements in the development of new landfill sites for a safe environment. In actual sites, the installment of an enhanced acquisition and removal system is considered based on data obtained by environment monitoring of the landfill site. Leachate movement and leachate levels within the filled areas are of particular importance for landfill practice. An uncontrolled outflow of leachate may adversely affect the surrounding environment and the aquatic system. Leachate collection involves two stages: an installation that leads leachate to a small number of collection points and the abstracting of leachate from the assembling points themselves. While it is feasible with pump systems for new cells or phases to be constructed in unfilled areas, the collection system should be installed together with wide-ranging basal leachate drainage blankets. These are also important at new containment landfills. Basal leachate drainage blankets consist of a series of pipes across the base of the site, which meet at different collection points. The collection pipe-work must be enclosed by at least 0.6 m (Pires et al., 2011) of low-fines aggregate and granular material. This granular layer supports the movement of leachate toward the collection system. This procedure could cause localized blockages of the leachate collection pipes, which are less likely to cause a problem in the operation of the drainage system. The base of the site must be established so that at least a 2:60 (Scheinberg, 2001) declivity is achieved with the oversight of the leachate collection points.

The effluvium media selected should

- Be structurally robust to withstand loading
- Be sufficiently coarse to preclude blockage
- Have a semi-permeability or minimum permeability of 1×10^{-3} m/s (Seadon et al. 2010)
- Not be sensitive to chemical incursion by the leachate

The sides of a landfill should be a gradual slope, and the blanket should be extended up the sides. This allows for proper installation of a rodding point from the surface to dislodge the blockage, so that the pipe works continuously. In the selection of collection pipes, an order should be made for their estimation under the load from the filled materials as above. In addition, as the filling progresses, leachate collection pipe work should be extended upward through the deposited bio-waste. This can be achieved with a continuous "wall" of leachate drainage material extending vertically through the site to the surface. Alternatively, leachate drains can be installed horizontally at different levels, but with vertical leachate collection points (Turner and Powell, 1991). These options will assist the migration of leachate to the collection pipe work and that means blockages at lower levels will not cause the system as a whole to malfunction. In deep sites where leachate pumping is required, leachate collection has traditionally been undertaken by the construction of a leachate collection chamber, which consists of mane whole rings. The rings are usually cast from sulfate-resistant concrete and should be drilled in the side to assist percolation. They should not normally have access ladders within them, as this may encourage unauthorized entry. Care should be taken to ensure that the chambers have good foundations that will not subside when a succession of rings is placed on top of each other and that the weight of the rings will not damage any liner system. This may lead to the substitution of the concrete rings by artificial materials such as plastics. Vertical leachate collection chambers should be surrounded by a porous drainage media, not deposited wastes, to assist in vertical percolation of leachate to the chamber. Several configurations of leachate collections point are possible. The use of low-angled leachate raisers laid parallel to the side of the site should be considered. Although not suitable for sites with steep sides, the system exerts much lower pressures on the liner system. A second advantage is that vertical chimneys often suffer from sideways movement due to settlement. The small-angle riser system is less prone to damage from the filling process as they are located at the perimeter of the phase. Pumps can be introduced by way of a skid system. When a site is suitable, gravity drainage systems should be considered.

In addition, leachate extraction should be used; sub-miscible pumps and various leachate dewatering techniques have been well developed attached with utilize educator pumps. These have several advantages, including being easy to install in small-diameter boreholes (150 mm) (UNEP, 2012). Also, educators do not contain movable parts, while the only maintenance necessary is the cleaning of slime from the spray head. But, the primary purpose of leachate removal is that significant leachate heads should not be allowed to build up in any landfill in an uncontrolled fashion. Unless the surrounding land is equally saturated, leachate may escape from the site through the base of the sides. However, it could also cause significant pressure to occur on the inside of temporary structures. Bunds used to separate the filled and unfilled phases of the site are particularly vulnerable. Accordingly, it is desirable that a leachate head of no more than 1 m should occur and gravity-drained sites can achieve a significantly lower leachate level.

All landfills should be frequently monitored for leachate levels and the frequency should be decided according to the USEPA Landfill Monitoring Manual. But this kind of monitoring point should be located independently, without any leachate pumping chambers, while the location and number of control points are selected to give an exact picture of the leachate levels within the filled material. Because open chambers on the surface of the site are not favorable from a health and safety perspective, leachate collection chambers should have observable lids, which should only be left open when the pumping system is undergoing maintenance (Wang and Nie, 2001; Wilson et al., 2006).

5.6.3.3 Biogas Recovery from Landfill

Landfill gas is produced by the decomposition of biodegradable wastes present inside landfill sites. Gas composition varies according to the type of waste and the time that has elapsed since discharge within the site. The composition and quantity of the landfill gas emitted from a landfill are variable over both the short and long term. The typical composition of landfill gas is 35% carbon dioxide and 65% methane by volume (SEPA 2002). It consists of trace quantities of vapors and organic gases; some of the gases may be potentially harmful to humans and animals, and may produce malodorous and corrosive compounds with combustion. Also, the methane content is flammable and forms potentially explosive mixtures of gases in certain conditions. As a result, there are concerns about its unconstrained release and migration. It has a high calorific value and hence there is much current interest in burning landfill gas for power generation and process heating. The primary aim of landfill gas flaring is to control odor nuisance, health hazard, and adverse environmental impacts, and safe disposal of the flammable constituents.

Proper attention should be given to achieve the effective destruction of harmful gases while flaring; this will reduce the environmental impacts and health risks associated with combustion products (SEPA, 2002). Landfills are a potential source of greenhouse gases and contribute about 22% of methane emissions worldwide, with 48% of man-made methane emissions in the Unites States being derived from landfills in 1996. The global warming potential (GWP) of methane is between 21 and 62 (cf. carbon dioxide has GWP of 1), depending on the period considered (SEPA, 2002).

In any period, all phases of degradation can be found in a landfill due to the continuous accumulation of solid waste landfill. The rate of the decomposition process depends on local conditions such as climate, temperature, and waste composition. Anaerobic activities inhibit decomposition due to the presence of air inside the landfill; this results in the combustion of waste present inside the landfill and affects the generation and proportion of landfill gas. Also, variations in atmospheric pressure also influence the air content inside the landfill. The profile of landfill gas varies from place to place; around 50–200 components are found at trace-level concentration (SEPA, 2002). Approximately 350 minor constituents are estimated in landfill gas; most of them are organic in nature. Inorganic constituents include significant levels of hydrogen sulfide and trace amounts of ammonia mercury and volatile metallic compounds.

5.6.4 Carbon Storage in Landfill

Landfill gas mainly consists of carbon dioxide and methane, and contains several trace elements with potential health hazards. Some of these are potentially explosive, and some greenhouse gases can be used as fuel and are normally considered as waste products and an asphyxiant. In addition, recent technology involved in the collection and combustion of landfill gas in the United Kingdom is at a comparatively advanced stage (SEPA, 2002). The United Kingdom has developed this technology extensively and applied it over the years. The UK environment agency has also developed a reticulation technique that separates oxygen and carbon dioxide from methane (SEPA, 2002).

6

Kinetics of Waste Degradation

Poornima Jayasinghe and Patrick Hettiaratchi

6.1 Waste Degradation Process

Solid waste degradation involves many different physical, chemical, and biological processes and generates gaseous and dissolved compounds. When the waste is mostly organic, biological processes will dominate the waste degradation process. Solid waste degradation occurs under two conditions: aerobic waste degradation in the presence of oxygen by aerobic microorganisms, and anaerobic waste degradation in the absence of oxygen by anaerobic microorganisms. As shown in Figure 6.1, the aerobic biodegradation of solid waste produces carbon dioxide, leachate, biomass, and decomposed organic matter (compost), while anaerobic biodegradation leads to the generation of biogas, a mixture of approximately 45%–60% methane and 40%–60% carbon dioxide, leachate, and biomass (Mata-Alvarez 2003a and b).

The primary useful products of aerobic and anaerobic degradation of waste are compost and methane gas, respectively. This chapter provides a description of the aerobic composting process and two anaerobic processes, landfilling and anaerobic digestion, together with the process mechanisms, key factors, and reaction kinetics of the waste degradation and gas production processes.

6.2 Aerobic Waste Degradation Process

The primary organic compounds in municipal solid waste (MSW) are carbohydrates, proteins and amino acids, hydrocarbons, and some compounds resistant to biodegradation (Jakobson 1994). These compounds undergo many chemical and biochemical changes during aerobic composting. The end products are carbon dioxide, water, and partially degraded organic materials or biomass, which is also called *compost*. Aerobic microbes, primarily

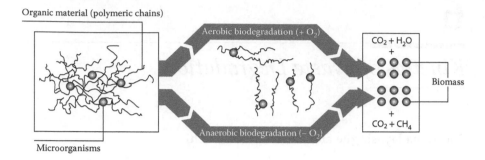

FIGURE 6.1
Aerobic and anaerobic waste biodegradation. (From API, APINAT BIO bioplastics, 2016. http://www.apinatbio.com/eng/apinat-bioplastics.html.)

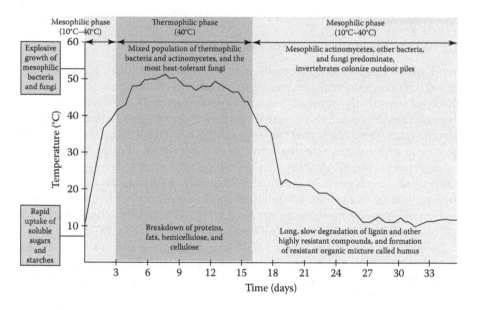

FIGURE 6.2
Three phases of composting. (From Trautmann and Krasny, *Composting in the Classroom*, pp. 1–26, Ithaca, NY: Cornell University, 1997.)

bacteria, archaea, and fungi, extract energy from the organic waste materials through a series of exothermic reactions that break the material into simpler molecules (Vesilind et al. 2002).

The aerobic composting process usually goes through three distinct phases, based on the temperature of the system. Figure 6.2 shows the three phases of composting: initial mesophilic, thermophilic, and secondary mesophilic.

6.2.1 Initial Mesophilic Phase

This is also called the *moderate temperature phase,* where the temperature rises to 40°C and typically lasts a couple of days. The initial decomposition of waste is carried out by mesophilic microorganisms. These microorganisms thrive at moderate temperatures and rapidly break down the soluble and readily degradable compounds in the waste. This is an exothermic process and the temperature of the system rises exponentially. Once the temperature exceeds 40°C, the mesophilic microorganisms become less competitive and thermophilic microorganisms become predominant.

6.2.2 Thermophilic Phase

This is a high-temperature phase where the temperature of the system increases over 40°C, which can last from a few days to several weeks, depending on the size of the system and the composition of the feed material. The high temperature of the system accelerates the breakdown of proteins, fats, and complex carbohydrates such as cellulose and hemicellulose.

6.2.3 Secondary Mesophilic Phase

When the system becomes substrate limited, the temperature of the system gradually decreases and mesophilic microorganisms start growing again. This is the final phase of the mesophilic curing or maturation of the remaining organic matter, and this phase proceeds for several months.

6.2.4 Stoichiometric Equation for Aerobic Waste Degradation

The general equation for aerobic conversion of waste can be presented as (Vesilind et al. 2002)

$$\text{Complex organics} + O_2 \xrightarrow{\text{aerobes}} \text{compost} + CO_2 + H_2O + NH_3 + SO_4^{-2} + \text{heat}$$

(6.1)

The aerobic waste degradation process can be stoichiometrically expressed as

$$C_aH_bO_cN_d + 0.5(ny + 2s + r - c)O_2 \rightarrow nC_wH_xO_yN_z$$

$$+ sCO_2 + rH_2O + (d - nz)NH_3$$

(6.2)

where:

$C_aH_bO_cN_d$ is the molecular formula for organic waste material

$C_wH_xO_yN_z$ is the molecular formula for the compost product

If complete conversion of the substrate is accomplished, Equation 6.2 will reduce to the following form:

$$C_aH_bO_cN_d + \left(a + \frac{b}{4} - \frac{c}{2} - \frac{3d}{4}\right)O_2 \rightarrow aCO_2 + \left(\frac{b}{2} - \frac{3d}{2}\right)H_2O + dNH_3 \quad (6.3)$$

If the chemical compositions of waste and the compost are known, the amount of compost produced and the air requirement can be theoretically calculated using Equation 6.2.

6.2.5 Factors Affecting Aerobic Degradation

Composting is not an instantaneous process, and many environmental and physical conditions affect this microbial process, both negatively and positively (Huag 1993). Some of the parameters that control the aerobic composting process are moisture content, feedstock composition, oxygen requirements, and pH.

6.2.5.1 Moisture

One of the major requirements of microbial processes is the availability of moisture. Insufficient moisture in the system may slow down the waste degradation rate and excess moisture may restrict the oxygen transfer, creating anaerobic pockets within the compost (Trautmann and Krasny 1997). An initial moisture content of 50%–65% by weight is considered optimum for composting (Trautmann and Krasny 1997; Weppen 2001).

6.2.5.2 Feedstock Composition (C:N ratio)

The carbon-to-nitrogen (C:N) ratio affects not only the performance of the composting process, but also the quality of the compost product. The ideal C:N ratio for composting is generally considered to be around 30:1 (Trutmann and Krasny 1997). The optimum ratio largely depends on the relative abundance of microbes present in the compost, as bacteria require more nitrogen than fungi. C:N ratios lower than the optimum value allow rapid degradation, but excess nitrogen will be converted to ammonia gas, causing undesirable odors and a loss of nitrogen from the compost product. C:N ratios higher than the optimum value do not provide sufficient nitrogen for the optimal growth of microbial populations, and may sufficiently inhibit population growth to prevent the temperature increasing to the thermophilic phase. As carbon is converted to carbon dioxide during the composting process, the C:N ratio decreases; the typical C:N ratio at the end of composting is around 15:1.

6.2.5.3 Oxygen Requirements

One of the essential requirements for aerobic microbes is the presence of oxygen, and the efficiency of an aerobic composting system is closely related to the aeration flow rate (Ahn et al. 2007). Insufficient aeration can lead to anaerobic conditions with undesirable odors, while too much air supply can lead to excessive cooling and impede the thermophilic conditions required for optimum rates of decomposition. Oxygen concentrations greater than 10% are considered optimal for aerobic composting (Trautmann and Krasny 1997; Ahn et al. 2007). Furthermore, maintaining the proper balance between moisture and oxygen is one of the key conditions for a successful composting operation.

6.2.5.4 pH

During the composting process, the pH generally varies between 5.5 and 8.5 (Trumann and Krasny 1997). In the early stages of composting, the accumulation of organic acids causes a decrease in pH, which encourages the growth of fungi. Further decomposition and volatilization of organic acids increases the pH of the system. Later in the composting process, the pH tends to become neutral and the final compost generally has a pH between 6 and 8.

6.3 Anaerobic Waste Degradation Process

Anaerobic waste degradation is a process by which biodegradable waste is decomposed by anaerobic microorganisms in the absence of oxygen, producing methane and carbon dioxide. Simplified theoretical anaerobic waste degradation pathways are presented in Figure 6.3.

6.3.1 Hydrolysis Stage

The first stage of anaerobic waste degradation is hydrolysis, and it is a very important process in anaerobic waste degradation, since the solid organic waste must be solubilized before the microorganisms can utilize it. In this stage, complex organic compounds such as proteins, carbohydrates, and lipids are broken down into simple organic matter such as amino acids, sugars, and fatty acids, respectively, by the action of extracellular enzymes excreted by hydrolytic bacteria. Some of the important chemical reactions that occur during the hydrolysis stage and their associated free-energy values are as follows (Mclnerney et al. 1979; Jones and Grainger 1983; Mata-Alvarez 2003a and b).

$$C_6H_{12}O_6 + 4H_2O \rightarrow 2CH_3COO^- + 2HCO_3^- + 4H^+ + 4H_2$$

$$\rightarrow (\Delta G^\circ = -206 \text{ kJ/reaction})$$

$$C_6H_{12}O_6 + 2H_2O \rightarrow 2CH_3CH_2OH + 2HCO_3^- + 2H^+ \rightarrow (\Delta G^\circ = -226 \text{ kJ/reaction})$$

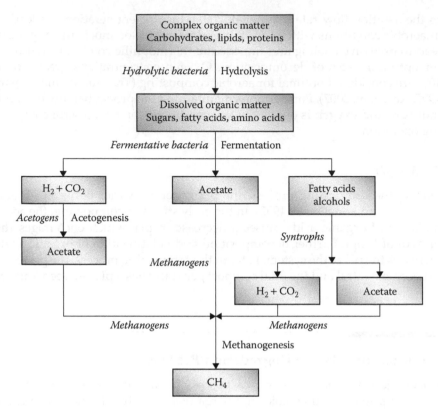

FIGURE 6.3
Anaerobic waste degradation pathways. (From Hettiaratchi et al., *Current Organic Chemistry*, 19(5), 413–422, 2015.)

$$C_6H_{12}O_6 + 2H_2 \rightarrow 2CH_3CH_2COO^- + 2H_2O + 2H^+$$

$$\rightarrow \left(\Delta G^\circ = -358 \text{ kJ/reaction}\right)$$

$$C_6H_{12}O_6 \rightarrow 2CHCHOHCOO^- + 2H^+ \rightarrow \left(\Delta G^\circ = -198 \text{ kJ/reaction}\right)$$

Hydrolysis is considered to be the rate-limiting step in the solid waste biodegradation process (Haarstrick et al. 2001; Garcia-Heras 2003).

6.3.2 Acetogenesis Stage

The second stage of anaerobic waste degradation is the acetogenesis stage, which can be explained in terms of two substages. In the initial fermentation substage, the products from the hydrolysis stage are further broken down by fermentative bacteria into simple molecules such as short-chain carboxylic acids, acetate, carbon dioxide, and hydrogen. In the acetogenesis substage, the simple molecules from fermentation are further digested by acetogens to produce hydrogen, carbon dioxide, and acetate. The fermentation of some intermediate products such as fatty acids (e.g., propionate and butyrate) to acetate, carbon dioxide, and hydrogen is thermodynamically favorable only at very low hydrogen concentrations. Therefore, for acetogenesis to remain energetically favorable, the methanogens must sufficiently scavenge the available hydrogen. This cooperative phenomenon is referred to as *syntrophy* (Zehnder 1978; Mormile et al. 1996; Barber 2007).

Some of the important reactions that occur in acetogenic processes and the free-energy changes associated with the given reaction under standard conditions are as follows (Khanal 2008).

$$CH_3CH_2COO^- + 3H_2O \rightarrow CH_3COO^- + H^+ + HCO_3^- + 3H_2$$

$$\rightarrow \left(\Delta G^\circ = +76.1 \ kJ/reaction \right)$$

$$CH_3CH_2CH_2COO^- + 2H_2O \rightarrow 2CH_3COO^- + H^+ + 2H_2$$

$$\rightarrow \left(\Delta G^\circ = +48.1 \ kJ/reaction \right)$$

$$C_7H_5CO_2 + 7H_2O \rightarrow 3CH_3COO^- + 3H^+ + HCO_3^-$$

$$+ 3H_2 \rightarrow \left(\Delta G^\circ = +53 \ kJ/reaction \right)$$

$$CH_3CH_2OH + H_2O \rightarrow CH_3COO^- + H^+ + 2H_2 \rightarrow \left(\Delta G^\circ = +9.6 \ kJ/reaction \right)$$

Because of the acid production and the effect of the elevated concentration of carbon dioxide during this stage, the pH of the leachate, if formed, will often drop to a value of 5 or lower (Christensen et al. 1989). A number of inorganic compounds, including heavy metals, and many essential nutrients will be solubilized during this stage due to the low pH value of leachate.

6.3.3 Methanogenesis Stage

In the methanogenesis stage, methane is produced by the methanogenic archaea. The two most common substrates for methanogens are hydrogen/carbon dioxide and acetate. In this stage, acetoclastic microbes convert acetate to methane, and hydrogenetrophic microbes convert hydrogen and carbon dioxide to methane.

Some of the important reactions involved in methanogenesis with associated energy yield are as follows (Khanal 2008).

$$4H_2 + CO_2 \rightarrow CH_4 + 2H_2O \rightarrow \left(\Delta G = -139 \ kJ/mol \right)$$

$$4HCOO^- + 2H^+ \rightarrow CH_4 + CO_2 + 2HCO_3^- \rightarrow \left(\Delta G = -127 \text{ kJ/mol} \right)$$

$$CH_3COO^- + H_2O \rightarrow CH_4 + 2HCO_3^- \rightarrow \left(\Delta G = -28 \text{ kJ/mol} \right)$$

$$4CH_3OH \rightarrow 3CH_4 + CO_2 + 2H_2O \rightarrow \left(\Delta G = -103 \text{ kJ/mol} \right)$$

$$4CH_3NH_2 + 2H_2O + 4H^+ \rightarrow 3CH_4 + CO_2 + NH_4^+ \rightarrow \left(\Delta G = -102 \text{ kJ/mol} \right)$$

$$\left(CH_3 \right)_2 S + H_2O \rightarrow 1.5CH_4 + 0.5CO_2 + H_2S \rightarrow \left(\Delta G = -74 \text{ kJ/mol} \right)$$

The conversion of acetic acid to methane is the primary pathway for methane production, and also raises the threshold of the system to more neutral values in the range of 6.8–8.0 (Christensen et al. 1989).

6.3.4 Stoichiometric Equation for Anaerobic Waste Degradation and Estimation of Theoretical Methane Yield

The generalized equation for anaerobic organic matter bioconversion is expressed as (Vesilind et al. 2002)

$$\text{Complex organics} + H_2O \xrightarrow{\text{anaerobes}} CH_4 + CO_2$$

$$+ NH_4^+ + H_2S + \text{biomass} + \text{heat} \tag{6.4}$$

If the chemical composition of the waste is known, the theoretical methane yield can be predicted using the Buswell equation (Mata-Alvarez 2003a and b). The overall process of converting organic compounds to methane and carbon dioxide may stoichiometrically be expressed as

$$C_aH_bO_cN_dS_e + \left(a - \frac{b}{4} - \frac{c}{2} + \frac{3d}{4} + \frac{e}{2} \right)H_2O \rightarrow \left(\frac{a}{2} + \frac{b}{8} - \frac{c}{4} - \frac{3d}{8} - \frac{e}{4} \right)CH_4$$

$$+ \left(\frac{a}{2} - \frac{b}{8} + \frac{c}{4} + \frac{3d}{8} + \frac{e}{4} \right)CO_2 + dNH_3 + eH_2S \tag{6.5}$$

This equation can be applied only to the biodegradable volatile solid (BVS) fraction of waste. The molecular formula for solid waste can be estimated if the chemical compositions of the waste are known. An elemental analysis can provide an estimate of the chemical composition.

As an example, consider an organic fraction of waste (volatile solids; VS) with the molecular formula $C_{99}H_{149}O_{59}N$ (Vesilind et al. 2002). Assume that

Equation 6.5 can be applied directly to the VS of the waste. With the given molecular formula, the theoretical methane yield is calculated as 275 L/kg of VS of waste, assuming 50% of waste is biodegradable. The calculation for estimating the theoretical methane production can be presented in several steps:

Step 1: The balanced equation for the given waste (VS of waste) can be written as

$$C_{99}H_{149}O_{59}N + 33H_2O \rightarrow 53CH_4 + 46CO_2 + NH_3$$

Step 2: The molecular weights of individual components are known ($C = 12$ g/mol; $H = 1$ g/mol; $O = 16$ g/mol; $N = 14$ g/mol). The molecular weight of methane can be calculated as

$$(1 \text{ mol} \times 12 \text{ g/mol}) + (4 \text{ mol} \times 1 \text{ g/mol}) = 16 \text{ g}$$

   Similarly, the molecular weight of the given waste can be calculated as 2295 g.

Step 3: The theoretical methane yield can be estimated according to the balanced equation in Step 1:

1 mole of $C_{99}H_{149}O_{59}N$ produces 53 moles of CH_4.

1 g (or kg) of $C_{99}H_{149}O_{59}N$ produces $(53 \times 16)/2295 = 0.37$ g (or kg) of CH_4.

The density of methane is 0.67 kg/m³ at 20°C and 1 atm.

1 kg of $C_{99}H_{149}O_{59}N$ produces $(0.37/0.67) = 0.55$ m³ of CH_4.

   Assuming that only 50% of VS is biodegradable (BVS) in the given waste, 1 kg of VS of $C_{99}H_{149}O_{59}N$ produces 275 L of CH_4.

   Similarly, theoretical carbon dioxide (CO_2), ammonia (NH_3), and hydrogen sulfide (H_2S) production can be calculated using the corresponding stoichiometric coefficient.

6.3.5 Factors Affecting Anaerobic Waste Degradation

The anaerobic waste degradation process may fail if one group of microorganisms or one process is inhibited. Therefore, well-controlled environmental parameters need to be maintained throughout the process (Mata-Alvarez 2003a and b). The major factors affecting the anaerobic waste degradation process include pH, moisture content, temperature, nutrients, and inhibitors.

6.3.5.1 pH

Methanogens are more susceptible to pH variation than other microorganisms in the microbial community (Khanal 2008). Therefore, the anaerobic waste degradation process is affected by slight variations of pH from optimum. The optimum pH for acetogens is 5.5–6.5 and for methanogens is 6–8 (Jones and Grainger 1983; Barlaz et al. 1990); however, the optimum pH for the combined culture ranges from 6.8 to 7.4, with neutral pH being the ideal (Khanal 2008).

6.3.5.2 Moisture Content

Adequate moisture is an essential requirement in a functional bioreactor. The availability of moisture promotes microbial degradation in two ways. First, it facilitates the movement of nutrients and microbes through solid waste and, second, it flushes out soluble pollutants and degradation products. The methane production rate increases as the moisture content of the waste increases. Research findings suggest that an exponential increase in the gas production rate occurs between 35% and 60% moisture content (Barlaz et al. 1990; Hamoda et al. 1998).

6.3.5.3 Temperature

The anaerobic waste degradation process is affected by temperature from both a kinetic and a thermodynamic point of view; both the rate and yield are directly dependent on temperature (Mata-Alvarez 2003a and b). Methanogen populations in waste are typically composed of a mesophilic group with a maximum rate of gas production at around 35°C–40°C, and a thermophilic group with a maximum around 55°C (Mata-Alvarez 2003a and b; Khanal 2008); however, anaerobic digestion can function over a large range of temperatures from around 10°C–65°C (Scherer et al. 2001; Sternenfels 2012).

6.3.5.4 Nutrients

In addition to the organic carbon substrate, microbes require a variety of macronutrients, such as nitrogen and phosphorous, and micronutrients, such as calcium, magnesium, potassium, nickel, iron, zinc, copper, cobalt, and some vitamins, for optimal anaerobic waste degradation. Various researchers have suggested different ratios of carbon (expressed as *chemical oxygen demand*, or COD), nitrogen (N), and phosphorus (P) based on the biodegradability of the waste, ranging from a COD:N:P ratio of 100:0.44:0.08 (McCarty 1964; Christensen and Kjeldsen 1989) to 350:7:1 (Khanal 2008); however, an average COD:N:P ratio of around 100:1.2:0.2 is recommended for a substrate to be efficiently anaerobically degraded (Mata-Alvarez 2003a and b).

6.3.5.5 Inhibitors

The absence of oxygen is an essential condition for the growth of anaerobic bacteria. Several other substances also exhibit inhibitory effects when present at elevated concentrations, including ammonia nitrogen above 1500 mg/L (Khanal 2008) and hydrogen sulfide above 200 mg/L (Mata-Alvarez 2003a and b). However, some substances can act as either a stimulant or inhabitant depending on their concentration (McCarty 1964). For example, sodium is a stimulant when the concentration ranges between 100 and 200 mg/L, while it is an inhibitor when the concentration is higher than 3000 mg/L.

6.4 Waste Degradation Sequence in Landfills

Although the waste degradation process in landfills is complex, researchers have developed a theoretical or idealized sequence of anaerobic waste degradation processes that occur in landfills by using the simplified waste degradation pathways presented in Section 6.3 and the experience gained from full-scale landfill operations. Figure 6.4 illustrates such an idealized sequence for a homogeneous volume of waste (Christensen et al. 1989). It involves five distinct phases.

Phase I is a short aerobic phase immediately after waste is placed in the landfill. In this stage, aerobic degradation occurs because a certain amount of air is trapped within the landfill. Oxygen may diffuse into the landfill waste from the atmosphere, and aerobic bacteria in the top layers of the landfill will readily consume the oxygen and limit the aerobic zone. This is also called an *acclimation period*, where the microbial community adapts to the newly anoxic environment within the waste mass. The easily degradable organic fraction of the solid waste is aerobically degraded during this phase and generates carbon dioxide.

Phase II is a transition phase, during which oxygen is depleted and anaerobic conditions begin to develop. The activity of the fermentative and

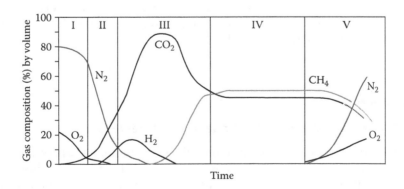

FIGURE 6.4
Waste degradation sequence.

acetogenic bacteria results in the rapid generation of carbon dioxide and some hydrogen. By the end of this phase, the COD of the leachate ranges between 480 and 18,000 mg/L, depending on waste characteristics and other environmental conditions (Vesilind et al. 2002).

Phase III is also an intermediate anaerobic phase. The hydrolysis of solid waste continues in this stage, followed by the microbial conversion of biodegradable organic content. Slow growth of methanogenic bacteria can occur and the methane concentration in the gas increases, while hydrogen and carbon dioxide concentrations decrease. A decrease in pH is often observed and the COD in the leachate ranges between 1500 and 71,000 mg/L (Vesilind et al. 2002).

Phase IV is the methanogenesis phase, where intermediate acids are consumed by methanogens and converted into methane and carbon dioxide. In this phase, a fairly stable methane production rate results in a methane concentration in the biogas of 50%–65% by volume. The pH increases as a result of bicarbonate production and the COD in the leachate is decreased to a range of 580–9760 mg/L.

Phase V is known as the maturation phase. It occurs after the readily available biodegradable organic material has been converted to methane and carbon dioxide in the earlier phases. In this phase, the rate of biogas generation diminishes significantly because of substrate and/or nutrient limitations. The leachate strength stays steady at much lower COD concentrations between 31 and 900 mg/L. Nitrogen and oxygen will start appearing in the landfill gas again due to increased diffusion from the atmosphere as a result of decreased internal pressure. However, the slow degradation of resistant organic fraction may continue with the production of humic-like substances.

Conditions within an actual landfill will differ considerably from this simplified sequence for a homogeneous mass of waste. The heterogeneous nature of landfilled waste will result in separate areas progressing through these stages at varying rates (if at all), considerably changing the overall trends for the landfill as a whole. Progress toward final solid waste stabilization depends on the environmental conditions, the age and characteristics of the waste, the operational and management conditions, and the site-specific external conditions (Vesilind et al. 2002).

6.5 Waste Degradation in Anaerobic Digesters

Anaerobic digester (AD) technology uses the same biological anaerobic waste degradation process that occurs within a landfill; however, this occurs within a controlled system where operational parameters can be regulated for optimal conditions, resulting in improved biogas production rates and volumes. Several different modes of AD system are available, depending on which factors are controlled in the system (Mata-Alvarez 2003b; Ward et al. 2008). AD systems are categorized as follows:

6.5.1 Moisture Controlled

Depending on the total solids concentration of the feed substrate, AD systems can be categorized as wet or dry. Wet AD systems are designed to process a feedstock with <15% of total solids and is commonly used to process waste such as sewage sludge and food industry effluents. Dry AD systems are used to digest waste consisting of around 35% total solids (Mata-Alvarez 2003a and b). Achieving proper mixing of microorganisms and substrate is more difficult in dry AD systems compared with wet AD systems, resulting in different rates of biogas generation.

6.5.2 Temperature Controlled

AD systems can be operated within psychrophilic (below 20°C), mesophilic (between 20°C and 40°C), or thermophilic (between 50°C and 65°C) temperature ranges. Because of the accelerated biochemical reactions, the waste degradation and biogas generation rate is highest under thermophilic conditions (Monnet 2003).

6.5.3 Feedstock/Substrate Delivery Controlled

AD systems can be operated as batch, single, or multistage continuous and semicontinuous reactors depending on the method of substrate feeding (Ward et al. 2008). In the batch system, digesters are filled with feedstock, with or without the addition of an inoculum, and no more additions take place during its operation. In continuous AD systems, the feedstock is fed and discharged from the digestion reactor continuously, with biogas generation occurring at a reasonably constant rate. Semicontinuous digesters are fed at continuous intervals of time, with simultaneous removal of the content.

Two other factors that determine the performance of the AD system are solids retention time (SRT) and mixing. SRT determines the treatment capacity of the digester and values vary with other factors, including temperature and waste composition. Generally, SRT ranges from less than 21 days to 240 days. A high retention time increases the biogas production from a set amount of feedstock due to better stabilization (Monnet 2003; Kim and Oh 2011). Mixing is also a critical parameter; adequate mixing can enhance biogas production due to the distribution of substrates, enzymes, and microorganisms throughout the digester (Kalia and Singh 2001). Furthermore, mixing prevents scum formation and avoids temperature gradients within the digester.

6.6 Waste Degradation Reaction Kinetics

Studying the kinetics of solid waste biodegradation is important for two main reasons (Wang 2004; Qdais and Alsheraideh 2008):

- The practical importance of being able to predict how long it takes to achieve the stabilization of solid waste.
- The study of reaction rates leads to an understanding of the mechanisms of the reaction.

Compared with the aerobic waste degradation process, the anaerobic waste degradation process consists of simultaneous reactions, reactants, and products, and involves different groups of bacteria. This makes it difficult to develop a quantitative measure of the transformation phenomena. The level of complexity of any waste degradation model can vary widely depending on the number of variables included in the model and the simplified assumptions that they make. The following sections represent simplified models used by many researchers (Heldivakis 1983; El-Fadel et al. 1988; Jayasinghe 2013).

6.6.1 Anaerobic Waste Degradation: Kinetic Model

As described in previous sections and shown in Figure 6.5, a waste cell consists of solid, liquid, and gas phases. The solid phase includes organic solid waste, inorganic matter, and nondegradable inert material. Organic solid waste can be divided into three fractions according to the degree of

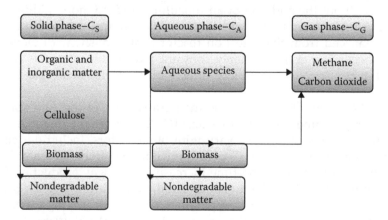

FIGURE 6.5
Simplified carbon degradation pathways. (Adapted from Jayasinghe, Enhancing gas production in landfill bioreactors by leachate augmentation, PhD thesis, Department of Chemical and Petroleum Engineering, University of Calgary, Canada, 2013.)

biodegradability: cellulose, hemicellulose, and lignin. However, each of these fractions are eventually converted to gas. The solid first dissolves into the leachate by enzymatic hydrolysis. In the aqueous phase, various aqueous species such as acetate and water-soluble monosaccharides are in equilibrium. The species in the aqueous phase are available to the biomass for conversion into gas. Biomass is a coexisting component in the solid and aqueous phases. Biomass can undergo self-decay or eventually can convert into gas. The gas phase mainly consists of methane and carbon dioxide.

The substrate that is used for modeling is generally carbon (El-Fadel et al. 1988): solid phase carbon (C_S), aqueous phase carbon (C_A), and gas phase carbon (C_G).

Furthermore, when formulating the model, any inhibitory effects are neglected. The growth and decay of microbial mass are not accounted for in this model since the model pathway assumes that the microbial mass will eventually be mineralized to gas (methane and carbon dioxide).

6.6.1.1 Solid Hydrolysis

The hydrolysis of solid substrate plays an important role in the waste degradation process. There are two approaches to estimating the solid hydrolysis kinetics:

- Using first-order waste hydrolysis kinetics
- Using naturally occurring enzyme-catalyzed kinetics

6.6.1.1.1 Method 1: First-Order Waste Hydrolysis Kinetics

Most of the research suggests that first-order kinetics with respect to substrate is the most suitable way to express solid waste degradation hydrolysis (Hamoda et al. 1998; Bizukojc et al. 2002; Vavilin 2008; Mata-Alvarez 2003a and b). In this case, the limiting factor is the remaining amount of substrate or the amount of methane already produced; other factors such as moisture or availability of nutrients are not supposed to be limiting factors. Although it is clear that these factors influence methane production, the suggested first-order kinetics appears to be supported by the fact that methane production gradually declines in the long term.

The conversion of solid-phase carbon can be represented as

$$\frac{dC_S}{dt} = -K_h C_S \tag{6.6}$$

where K_h is the hydrolysis rate constant of solid waste that is independent of concentration but dependent on temperature.

To find K_h, integrate the preceding equation between the initial waste concentration, C_{So}, and the concentration of waste at time t, C_{St}:

$$\int_{[C_{So}]}^{[C_{St}]} \frac{d[C_S]}{[C_S]} = \int_{to}^{t} -K_h\, dt \tag{6.7}$$

$$\left(\ln \frac{[C_{St}]}{[C_{So}]} \right) = -K_h\left(t_t - t_o\right) \tag{6.8}$$

where $[C_{So}]$ and $[C_{St}]$ are the initial concentration of substrate at time t_o and the concentration of substrate at time t_t, respectively. If measurable experimentally, the slope of the straight line of $\left(\ln[C_{St}]/[C_{So}]\right)$ vs. the $\left(t_t - t_o\right)$ plot is equal to the rate constant, K_h.

6.6.1.1.2 Method 2: Enzyme-Catalyzed Kinetics

The other approach to finding hydrolysis waste kinetics is that solid waste hydrolysis is assumed to be dominated by naturally occurring enzymes; hence, the overall rate of waste degradation reaction depends on the catalytic activity of the naturally occurring enzymes in the waste hydrolysis reaction. The kinetics of the general enzyme-catalyzed reaction may be simple or complex, depending on the enzyme and substrate concentrations, the presence/absence of inhibitors and cofactors, temperature, and pH. The simplest form of the rate mechanism for enzyme reactions was proposed by Michaelis and Menten in 1913, and is known as the Michaelis–Menten model. For simplification, it is assumed that enzyme-catalyzed reactions follow the mechanism shown as follows:

$$E + S \underset{k_2}{\overset{k_1}{\rightleftharpoons}} ES \xrightarrow{k_3} E + P \tag{6.9}$$

A single molecule of an enzyme (E) combines reversibly with a single molecule of substrate (S) to form an enzyme–substrate complex (ES), with the irreversible decomposition of the complex to a final product (P) and the release of a free enzyme (E). Rate constants k_1, k_2, and k_3 represent the rate of the reactions.

The Michaelis–Menten equation allows the rate of the enzyme-catalyzed reaction to be calculated as

$$v = -\frac{d[C_S]}{dt} = \frac{V_{max}[C_S]}{K_M + [C_S]} \tag{6.10}$$

where V_{max} is the maximum rate of reaction (velocity) and K_M is the Michaelis constant, $K_M = \left((k_2 + k_3)/k_1\right)$.

At low substrate concentrations ($[C_S] \lll K_M$), the rate is proportional to the substrate concentration; hence, the reaction is first order. At high substrate concentrations ($[C_S] \ggg K_M$), the rate of the reaction reaches V_{max} and becomes a zero-order reaction, independent of substrate concentration. Initially, in practice, the rate of the reaction is only restricted by the ability of the enzymes to utilize the substrate, which is in excess; thus, the reaction kinetics are zero order. However, as the substrate is utilized, the reaction begins to become substrate limited, resulting in fractional-order reactions until the substrate concentration is so low that the rate of reaction becomes totally limited by the substrate concentration, and, thus, first-order kinetics results.

The nonlinear form of the Michaelis–Menten equation does not permit the simple estimation of kinetic parameters. K_M and V_{max} can be determined by plotting the experimental results of the reaction rate and substrate concentration in one of the three different approaches that follow (Missen et al. 1999):

1. Use of initial rate data with a linearized form of the rate law; the most popular forms are as follows (Leskovac 2003):
   a. Lineweaver–Burk plot: $1/v = 1/V_{max} + K_M/V_{max}(1/[C_S])$

      $V_{max} = 1/\text{intercept}$ and $K_M = V_{max} \times$ slope of the plot of $1/v$ versus $1/[C_S]$
   b. Eadie and Hofstee plot: $v = V_{max} - K_M(v/[C_S])$

      $V_{max} =$ intercept and $K_M =$ slope of the plot of v versus $v/[C_S]$
   c. Hanes plot: $[C_S]/v = K_M/V_{max} + 1/V_{max}[C_S]$

      $V_{max} = 1/\text{slope}$ and $K_M =$ intercept $\times V_{max}$ of the plot of $[C_S]/[v]$ versus $[C_S]$
2. Use of concentration time data with a linearized form of the integrated rate law as follows:

$$\frac{[\ln C_S / C_{So}]}{C_{So} - C_S} = \frac{1}{K_M} - \frac{V_{max}}{K_M}\frac{t}{C_{So} - C_S}$$

   $K_M = 1/\text{intercept}$ and $K_M = V_{max}/\text{slope}$ of the plot of $[\ln C_S/C_{So}]/C_{So} - C_s$ versus $t/C_{So} - C_S$
3. Use of concentration time data with the integrated rate law in a non-linear form. This is more complex than the linearized approaches; however, accurate results can be obtained. To determine V_{max} and K_M, experimental data for $[C_S]$ versus t is compared with the values of $[C_S]$ predicted by the numerical integration of Equation 6.10; estimates of V_{max} and K_M are subsequently adjusted until the sum of the squared residuals is minimized (Missen et al. 1999).

In practice, it may be difficult to measure the intermediate product concentrations and the naturally occurring enzyme concentrations, so determining individual rate constants with this process can be challenging.

6.6.1.2 Intermediate Waste Degradation: Aqueous Carbon

Aqueous carbon is produced during solid carbon hydrolysis and is consumed during its conversion to gas, as follows. For simplicity, all transformation processes involved in acetogenesis and acidogenesis have been lumped together as a single step in the model.

Aqueous (water-soluble) carbon:

$$\frac{d[C_A]}{dt} = \left[\frac{d[C_A]}{dt}\right]_{\text{formation}} - \left[\frac{d[C_A]}{dt}\right]_{\text{consumption}} \tag{6.11}$$

where $[C_A]$ is the aqueous organic carbon concentration.

For first-order reactions:

$$\left[\frac{d[C_A]}{dt}\right]_{\text{formation}} = [Y_{S-A} K_h C_S] \tag{6.12}$$

For enzyme-catalyzed reactions:

$$\left[\frac{d[C_A]}{dt}\right]_{\text{formation}} = \left[Y_{S-A} \frac{V_{\max}[C_S]}{K_M + [C_S]}\right] \tag{6.13}$$

where Y_{S-A} is the dimensionless yield coefficient of converting solid carbon to aqueous carbon (aqueous carbon formed per mass of solid carbon consumed).

Aqueous carbon consumption, $[d[C_A]/dt]_{\text{consumption}}$, can be related to the net microbial growth using Monod kinetics, as described in the following section.

The concentration of aqueous or soluble carbon can be measured experimentally as COD or *volatile fatty acids* (VFA) (Sanders et al. 2003). The yield coefficients can be determined experimentally or estimated from the published values of previous researchers.

6.6.1.2.1 Kinetics of Bacterial Growth and Decay

The conversion rate of aqueous carbon to gaseous carbon depends on the growth rate of the microbial population. The performance of a solid waste degradation process can also be measured by the rate at which microorganisms metabolize the waste, which, in turn, is directly related to their rate of growth (Smith 1981).

When the energy, carbon source, and other environmental requirements for growth are fulfilled, the increase in biomass concentration (C_X) with regard to time is proportional to the biomass concentration itself (Stanier et al. 1986). The proportionality constant (μ) is called the *specific growth rate* of the biomass. The mathematical formulation of biomass growth kinetics starts from the following equation:

$$\frac{dC_X}{dt} = \mu C_X \tag{6.14}$$

where:
C_X is the biomass concentration
t is the time

The most commonly used model, relating microbial growth to substrate utilization, is the Monod equation (Smith 1981; Wang 2004). The Monod equation describes the relationship between the concentration of the substrate (aqueous carbon in this case, which is C_A) and the specific growth rate of biomass (μ). It is expressed as

$$\mu = \mu_{max} \frac{C_A}{K_s + C_A} \tag{6.15}$$

where:
μ_{max} is the maximum specific growth rate
K_s is the half-saturation constant, the substrate concentration at which the specific growth rate is one-half of μ_{max}

This equation demonstrates that the overall rate of metabolism is controlled by the substrate concentration, and is most rigorously applicable to soluble substrates under conditions where the concentration of microorganisms involved can be evaluated (Smith 1981).

On the other hand, the mass of microorganisms produced is related to the aqueous carbon consumption:

$$\frac{dC_X}{dt} = -Y_{A-X} \frac{dC_A}{dt} \tag{6.16}$$

where Y_{A-X} is the dimensionless yield coefficient of converting aqueous carbon to biomass.

The living microorganisms in a culture have a limited life span, after which they die; their death or decay releases substances that are incorporated into the biodegradation cycle (Gracis-Heras 2003). In relating the microbial change to the substrate utilization or the gas production, the net rate of microbial change needs to be accounted for, considering both the microbial growth and the decay rate, as follows:

$$\frac{dC_X}{dt} = (\mu - k_d)C_X \tag{6.17}$$

where:

k_d is the specific decay rate, which includes respiration and death
C_X is the microbial population size

Substituting the microbial change to the aqueous carbon consumption:

$$\left[\frac{dC_A}{dt}\right]_{\text{consumption}} = -\frac{1}{Y_{A-X}}(\mu_{\max}\frac{C_A}{K_s+C_A} - k_d)C_X \qquad (6.18)$$

In the initial period of waste placement that is at high substrate concentrations, $K_s + C_A \approx C_A$. Hence, the rate of the reaction becomes independent of substrate concentration; therefore, the reaction is zero order.

$$\left[\frac{dC_A}{dt}\right]_{\text{consumption}} = -\frac{1}{Y_{A-X}}(\mu_{\max} - k_d)C_X \qquad (6.19)$$

As the substrate is utilized and substrate concentrations are reduced, $K_s + C_A \approx K_s$, the rate is proportional to the substrate concentration; hence, the reaction is first order.

$$\left[\frac{dC_A}{dt}\right]_{\text{consumption}} = -\frac{1}{Y_{A-X}}\left(\frac{\mu_{\max}C_A}{K_s} - k_d\right)C_X \qquad (6.20)$$

For a given species of microorganisms, Y_{A-X}, $\mu_{\max}$, k_d, and K_s are constants.

If the biomass concentration over time is measured experimentally, $\mu_{\max}$ and K_s can be estimated by linearizing the Monod equation similar to the Michaelis–Menten method, as explained in Method 2 in Section 6.6.1.1.2. The yield coefficients can be determined experimentally or can be approximated from the published values of previous researchers.

The microbial growth rate is not only dependent on substrate and microbial concentrations, but it also depends on other factors. One of the most relevant factors is inhibition; this can be substrate inhibition or product inhibition (Gracis-Heras 2003). The simplified Monod kinetics explained previously can be modified to account for these inhibitions as well; however, the complexity of the model will increase.

6.6.1.3 Formation of Methane

The formation of methane from aqueous carbon is mathematically described as

$$\frac{d[C_G]}{dt} = Y_{A-G}\left[\frac{d[C_A]}{dt}\right]_{\text{consumption}} \qquad (6.21)$$

where:

[C_G] is the gaseous (methane) carbon concentration

Y_{A-G} is the yield coefficient of converting aqueous carbon to methane carbon

The actual gaseous methane production can be measured throughout the process and the yield coefficient can be estimated based on actual methane production.

The waste degradation models have to be calibrated for every scenario and all the parameters need to be determined experimentally. Regardless of the model used, the accuracy of the inputs drives the results and, therefore, it is critical to use highly accurate site-specific data.

6.7 Aerobic Waste Degradation: Kinetic Model

Compared with the anaerobic reaction kinetic models, the aerobic reaction kinetic model does not involve many intermediate stages. Researchers have used first-order reaction kinetics for the initial hydrolysis stage of aerobic degradation (El-Fadel et al. 1996; Slezak et al. 2012). The hydrolysis stage can be expressed similarly to the hydrolysis stage of anaerobic degradation as illustrated in Equation 6.6 (Section 6.6.1.1).

Equation 6.22 describes the rate of change in organic carbon in leachate. It includes the formation of organic carbon in leachate due to hydrolysis, organic carbon decomposition by microorganisms in the leachate, and the organic carbon content from dead microorganisms in the leachate (Slezak et al. 2012).

$$\frac{d[C_l]}{dt} = K_h C_S - \mu_{max} \frac{C_l}{K_S + C_l} C_X + K_d C_X \tag{6.22}$$

where C_l is the organic carbon content of leachate during aeration and all other parameters are as explained in Section 6.6.1.2.

For simplification, it is assumed that the all carbon from dead microorganisms decomposes completely to carbon dioxide (El-Fadel et al. 1996). The rate of carbon dioxide production can be written as

$$\frac{d[C_{CO_2}]}{dt} = (1 - Y_a)\mu_{max} \frac{C_l}{K_S + C_l} C_X \tag{6.23}$$

where Y_a is the yield coefficient determining the amount of organic carbon in leachate used for microorganism growth.

Unlike the anaerobic process, the depletion or the assimilation of oxygen is also a factor in the aerobic process, and can be modeled as (Lin et al. 2008)

$$\frac{d[C_{O_2}]}{dt} = \mu_o \frac{C_A}{K_S + C_A} C_X \qquad (6.24)$$

where μ_o is the maximum specific rate of oxygen assimilation by microorganisms.

6.8 Landfill Gas Generation Kinetics

The majority of reactions taking place in waste degradation landfills are catalyzed by naturally occurring microorganisms. As described in Section 6.6, a common practice for measuring the performance of the solid waste degradation process is to measure the rate at which the microorganisms metabolize the waste, which can be related to their rate of growth and gas production (Wang 2004).

However, several other models are available for estimating the rate of landfill gas generation. A mass balance approach is the simplest emission estimation method, using the stoichiometric equation described in Section 6.3.4; however, it generally overestimates emissions. Other useful models for landfill emissions estimation are the Scholl Canyon model, the US Environmental Protection Agency (USEPA) Landfill Gas Emissions Model (LandGEM), the Intergovernmental Panel on Climate Change (IPCC) First-Order Decay (FOD) model, the IPCC default method, and the triangular method (Kumar et al. 2004; USEPA 2005; IPCC 2006). These models vary widely, not only in the assumptions that they make, but also in their complexity and the amount of data they require.

6.8.1 Scholl Canyon Model and USEPA LandGEM

The model most commonly used by landfill practitioners is the Scholl Canyon model (USEPA 2005). The Scholl Canyon model assumes that, after a lag time of negligible duration, during which anaerobic conditions are established and the microbial biomass is built up and stabilized, the gas production rate is at its peak. The model also follows first-order reaction kinetics and does not consider any limiting factors such as moisture or nutrients.

The derivation of this model, for a given waste mass in place, is as follows (Emcon Associates 1980; USEPA 2005):

$$\frac{dL}{dt} = -kL \tag{6.25}$$

where:

  t is the time
  L is the volume of methane remaining to be produced after time t
  k is the gas production rate constant

Integration of the preceding equation gives the following expression for landfill gas production:

$$L = L_o e^{-kt} \tag{6.26}$$

where L_o is the total volume of methane ultimately to be produced at $t = 0$.
  For the methane production rate:

$$\frac{dG}{dt} = -\frac{dL}{dt} = kL = kL_o e^{-kt} \tag{6.27}$$

where:

  G is the total volume of methane produced prior to time t
  kL_o is the peak generation rate that occurs at time zero in units of volume per time

The Scholl Canyon model applies the preceding formula to the mass of waste landfilled during independent discrete periods in the active life of a landfill (e.g., the mass landfilled in a given month or year). The total generation rate is the summation of the generation rates of these discrete masses, and can be represented as

$$Q = \sum_{i=1}^{n} r_i k_i L_{oi} e^{-k_i t_i} \tag{6.28}$$

where:

  Q is the methane generation rate in the year of the calculation
  n is the number of years of waste placement
  r_i is the fraction of total refuse in submass i
  k_i is the gas generation rate constant for submass i in reciprocal time
  L_{oi} is the volume of methane ultimately to be produced at $t = 0$ for submass i
  t_i is the age in years of the waste section placed in the ith year

LandGEM is software that was developed by the USEPA (2005) for quantifying landfill gas emissions. LandGEM is based on the first-order decomposition rate equation and the Scholl Canyon model. The LandGEM model provides an automated tool for the estimation of landfill gas generation from a landfill cell, whether operated as a dry-tomb landfill or a landfill bioreactor.

The model allows the estimation of annual generation over a specified time period, and is described as follows (USEPA 2005):

$$Q = 2\sum_{i=1}^{n} kL_oM_ie^{-kt_i} \qquad (6.29)$$

where:
- Q is the total landfill gas emission rate (Mg/year)
- n is the number of years of waste placement
- M_i is the mass of solid waste placed in year i (Mg)
- L_o is the methane generation potential of the waste (m³/Mg)
- k is the methane generation rate (per year)
- t_i is the age in years of the waste section placed in year i
- i is 1 year

LandGEM is considered a screening tool; the accuracy of the output estimates is reliant on the accuracy of the data inputs. There are two key input parameters associated with this model: k and L_o. They are described as follows.

The methane generation potential L_o is the total amount of methane that a unit mass of refuse will produce when given sufficient time. It depends on the composition of the waste and can be calculated using the stoichiometric equation for anaerobic waste degradation (Equation 6.5).

The k value represents an overall rate constant and is indicative of the fraction of the waste that undergoes decomposition in the given year to produce methane. The k value for a given waste mass is related to its *half-life*—the time taken by the waste to decay to half of its initial mass. For first-order reactions, k can be determined using Equation 6.30:

$$k = \frac{0.693}{t_{1/2}} \qquad (6.30)$$

where $t_{1/2}$ is the half-life.

Determining the reaction rate constant k is very important for modeling landfills, but can be a very challenging task; k is affected by a number of factors including the waste composition, climatic conditions at the site, characteristics of the disposal site, and waste disposal practices. There are two approaches to selecting the half-life (or k value) for the calculation. The first approach assumes the degradation of different types of waste to be completely dependent on each other (the *bulk waste* option, in which a single value of k is chosen for the entire waste matrix). The second approach assumes that the degradation of different types of waste is independent of each other (the *waste composition* option, in which a k value for each component of the waste stream is chosen for calculations).

6.8.2 IPCC First-Order Decay Model

The FOD model was developed by the IPCC and it assumes that the degradable organic carbon (DOC) in the waste decays slowly over a few decades, during which methane and carbon dioxide are formed (IPCC 2006). If conditions are constant, the rate of methane formation depends solely on the amount of carbon remaining in the waste (first-order kinetics). Therefore, methane emissions from waste deposited in a landfill are highest in the first few years after closure and gradually reduce over time. The FOD model for methane generation can be represented as follows (IPCC 1998):

CH_4 generation in year t (Gg/year)

$$= \sum_x [(A.k.MSW_T(x).MSW_F(x).L_0(x)).e^{-k(t-x)}] \qquad (6.31)$$

where:

t is the year of inventory
x is the years for which input data should be added
$A = (1-e^{-k})/k$ is the normalization factor that corrects the summation
k is the methane generation rate constant (1/year)
$MSW_T(x)$ is the total MSW generated in year x (Gg/year)
$MSW_F(x)$ is the fraction of MSW disposed at solid waste disposal site in year x (Gg/year)
$L_0(x)$ is the methane generation potential [$MCF(x).DOC(x).DOC_F.F.16/12$] (Gg CH_4/Gg waste)
$MCF(x)$ is the methane correction factor in year x (fraction)
$DOC(x)$ is the degradable organic carbon (DOC) in year x
DOC_F is the fraction of DOC dissimilated
F is the fraction by volume of methane in landfill gas
16/12   is the conversion from carbon to methane

The input parameters used in the FOD model are more or less similar to the LandGEM model, except that the IPCC has attempted to make the projections for methane emissions more realistic by incorporating various correction factors into the equation.

6.8.3 IPCC Default Method

The IPCC default model uses a simple mass balance calculation that estimates the amount of methane emitted from the solid waste disposal sites by assuming that all methane is released the same year the waste is disposed of (Kumar et al. 2004; IPCC 2006). Compared with the IPCC FOD model, the default model does not reflect the time variation in solid waste disposal and the degradation process. The default method is based on the following equation:

CH_4 emissions (Gg/year)

$$= (MSW_T . MSW_F . MCF . DOC . DOC_F . F . 16/12 - R).(1 - OX) \qquad (6.32)$$

where:

MSW_T is the total MSW generated (Gg/year)

MSW_F is the fraction of MSW disposed at the solid waste disposal site

MCF is the methane correction factor (fraction)

DOC is degradable organic carbon

DOC_F is the fraction of DOC dissimilated

F is the fraction of methane in landfill gas

16/12 is the conversion from carbon to methane

R is the recovered methane (Gg/year)

OX is the oxidation factor

The IPCC default model is simple compared with other models and emission calculations require only input of a limited set of parameters, for which the IPCC guidelines provide default values where country-specific quantities and data are not available (IPCC 2006).

6.8.4 Triangular Model

The FOD model requires detailed information on current and historic waste quantities and compositions, making it very difficult to use the FOD model when historical data is not available. The triangular method uses a modified FOD method to remove this limitation. The area of the triangle shown in Figure 6.6 is equal to the gas released over the period from the solid waste depositions (Kumar et al. 2004).

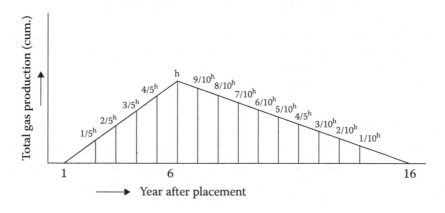

FIGURE 6.6

Triangular form for gas production. (Adapted from Kumar et al., *Atmospheric Environment*, 38, 3481–3487, 2004.)

In the absence of detailed data, the area of the triangle (volume of gas) is assumed to be equal to the volume estimated using the IPCC default method. In the triangular model, the gas production is analyzed under two different scenarios. The first phase starts after year 1 of the deposition of waste, when the gas production rate increases, and will continue for 6 years. The second phase starts when the gas generation rate decreases, and it is assumed that after 15 years the gas generation becomes zero.

6.9 Anaerobic Digester Kinetics

The anaerobic waste degradation model presented in Section 6.6.1 can be applied to AD systems as well, but is valid only for batch AD operations. Otherwise, the model needs to be modified based on the AD mode of operation. From a kinetics point of view, the reactor type illustrates different ways in which the rate of reaction can be measured experimentally and interpreted operationally (Missen et al. 1999). This section explains the development of anaerobic waste degradation kinetic equations based on different modes of operation: batch, continuous, and semicontinuous.

The general mass balance equation for a given control volume can be written as (Missen et al. 1999)

Rate of mass inflow − rate of mass outflow ± rate of generation/

consumption = rate of mass accumulation (6.33)

For a particular gas species (j), the general mass balance equation can be presented mathematically as

$$\frac{dN_j}{dt} = F_{j,\text{in}} - F_{j,\text{out}} + G_j \tag{6.34}$$

where:
N_j represents the number of moles of species j in the system at any time
$F_{j,\text{in}}$ is the mass flow rate entering the system (moles/time)
$F_{j,\text{out}}$ is the mass flow rate leaving the system (moles/time)
G_j is the net mass generation (or consumption, if negative) rate (moles/time)

The rate of generation of species j, G_j, is the product of the reaction volume, V, and the rate of formation of species j, r_j (moles/times, volume). Furthermore, assume that the rate of formation of j for the reaction varies with the position of the system volume. G_j can be presented as

$$G_j = r_j V = \int^V r_j dV \qquad (6.35)$$

6.9.1 Batch Digester

A batch digester is a closed system and the total mass of each batch is fixed. Since there is no inflow or outflow of the reactants or the products during the reaction being carried out, $F_{j,in} = F_{j,out} = 0$. The resulting mole balance equation for species j can be written as

$$\frac{dN_j}{dt} = \int^V r_j dV \qquad (6.36)$$

If the contents of the digester are perfectly mixed, there is no variation in the rate of reaction throughout the digester volume. For a constant volume batch digester, the mass balance equation can be written as

$$\frac{1}{V}\frac{dN_j}{dt} = \frac{d(N_j/V)}{dt} = \frac{dC_j}{dt} = r_j \qquad (6.37)$$

where C_j is the concentration of species j (moles/volume).

6.9.2 Continuous Flow Digester

In continuous flow digesters, the flow of both input and output streams through the digester is continuous. In general, when the digester is stirred continuously, it is referred to as a *continuous stirred tank reactor* (CSTR) (Missen et al. 1999; Davis and Davis 2003). Combining Equations 6.34 and 6.35, the general mole balance equation for a CSTR can be written as

$$\frac{dN_j}{dt} = F_{j,in} - F_{j,out} + \int^V r_j V \qquad (6.38)$$

For a CSTR operated at steady state, $dN_j/dt = 0$. Assume that there are no spatial variations in the rate of reaction and take the molar flow rate, F_j, as a product of the concentration of j (C_j) and the volumetric flow rate, Q (volume/time), which is represented as $F_j = C_j \times Q$.

The rate equation for a CSTR can be written as

$$r_j = \frac{C_{j,in}Q_{in} - C_{j,out}Q_{out}}{-V} \qquad (6.39)$$

6.9.3 Semibatch Digester

A semibatch digester is a variation of a batch digester in which reactants may be added or products may be removed intermittently or continuously as reactions proceed (Davis and Davis 2003). Assuming only reactants are fed continuously during the semibatch operation, the mole balance equation can be written as

$$\frac{dN_j}{dt} = C_{j,in}Q_{in} + \int^{V} r_j V \tag{6.40}$$

Compared with batch digesters, the gradual or intermittent addition of reactants and withdrawal of products can result in the higher conversion of reactants (Missen et al. 1999).

7

Systems Approaches in Municipal Solid Waste Management

7.1 Introduction

Solid waste management (SWM) systems are always developed with the objective of dealing with major concerns such as environmental, social, and economic issues, resource exploitation, and land use. Most developing countries around the world are facing the ever-growing problem of waste generation, treatment, and disposal. Rapid urbanization, weak institution capacities, and the absence of strong regulations are the primary reasons for the failure of most SWM systems. None of the SWM systems developed in high-, medium-, and low-income countries include all aspects of sustainability; also, they do not include all the relevant stakeholders, such as government officials, the public, and industry, in the decision-making and planning process (Morrisey and Browne 2004). There is a need to develop a broad systems perspective for SWM in developing countries where municipal corporations are unable to handle waste properly and to provide adequate facilities to ordinary people. For the development of a systems approach, first it is necessary to understand the various development *drivers* (factors that significantly impact the development of SWM) in developed countries and the challenges faced by developing countries. Knowledge of drivers assists in understanding the current scenario of SWM systems in developed as well as developing countries. This chapter focuses on the concept of integrated SWM from a systems perspective, followed by the various systems approaches to SWM and the need to solve the prolonged problem of SWM in developing countries. The later sections explain the principles of systems engineering, beginning with the definition of a system, systems boundaries, and types of approach to systems thinking. Different systems engineering methods are also described in detail to identify which method is suitable for a particular context.

7.2 Development Drivers for Solid Waste Management

For the development of any system, it is of utmost importance to study how waste management developed in the past and what the drivers are now. Knowledge about development drivers will help in designing sustainable SWM systems in developing and developed countries around the world (Wilson 2007). Figure 7.1 shows various development drivers for SWM.

7.2.1 Solid Waste Management Development in High-Income Countries

High-income nations developed their systems approaches for SWM in the nineteenth and twentieth centuries; not only governments but residents also understood the need for proper SWM. From various incidences occurring in past, it can be concluded that SWM systems have been affected by different drivers such as public health, environment, resource degradation, climate change, and public awareness (Marshall and Farahbakhsh

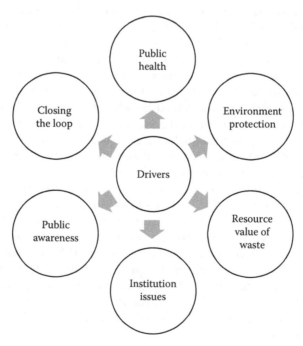

FIGURE 7.1
Development drivers for SWM.

2013). How these drivers shape SWM systems in high-income countries is explained as follows:

1. *Public health concerns*: In the late nineteenth and mid-twentieth centuries, most of the industrialized nations were deprived of basic sanitation facilities, and because of this, people suffered from widespread disease. During this period, governments in industrialized countries took various initiatives to collect and remove waste from underfoot so that the surroundings would be clean and diseases would not spread.

2. *Resource scarcity*: Before industrialization, resources were scarce, and the recycling and reuse of products were practiced in small industries as well as in homes. Every material was reused and repaired rather than sending it to landfill. When the industrialization period came, recycling practices increased with ragpickers and street buyers, who collected and sold waste to recycling industries for a living, and this practice is still continued in developing nations. Resources again increased in the 1970s, when there was a hike in resource consumption due to increases in population, increases in the marketing of commodities, and changes in living standards (Marshall and Farahbakhsh 2013). The concept of waste hierarchy came into existence in the late 1970s, which again sparked the recycling and reuse of waste material. Waste hierarchy has introduced various concepts such as reusing, reducing, recycling, energy recovery, treatment, and finally disposal at a landfill. The value of land as a resource was realized by governments when the size of landfills increased and there was a land shortage in small nations.

3. *Environmental concerns*: During the 1960s and 1970s, high-income countries recognized the need to conserve the environment and to reduce environmental pollution in their cities. At this time, the environmentalist movement started, which brought about a shift in approaches to waste management systems. Now, the focus was shifted from waste collection to waste disposal. Policy makers began making policies on disposal waste methods to prevent environmental contamination. During the 1970s and 1980s, policies were formed on converting open dump sites into covered and compacted landfills. From 1980 onward, high-income nations started creating environmental laws and standards for leachate control, incineration processes, dioxin reduction, and odor control (Marshall and Farahbakhsh 2013). After the 1990s, governments in high-income countries started focusing on the formation of integrated solid waste management (ISWM) systems, considering not only technical solutions but also social, economic, cultural, and political aspects too.

4. *Climate change*: Climate change has turned out to be a vital environmental driver, as many nations across the world understand the devastating impacts of climate change on their citizens. Until the end of the twentieth century, solid waste was disposed of in landfill, which is a major source of methane emissions, and this has led to the formation of many laws in developed nations for the conversion of open dump sites to sanitary landfills (UN-Habitat 2010).

5. *Public concerns and awareness*: In developed countries with increasing levels of living standards, people realized the need for proper SWM for the cleanliness of their surroundings and improvements in their health standards. But still, people in various developed nations have the notion of "not in my backyard," and local governments cannot plan for solid waste facilities within a city (Wilson 2007). Various SWM strategies such as reducing, reusing, recycling, and composting at home still need to be adopted by ordinary people; only then will the increasing burden on SWM facilities be reduced. A significant portion of the population is unable to make deliberate choices because they find themselves locked into unsustainable patterns caused by habits, daily routines, a lack of knowledge, inequality in access, social acceptance, and cultural values.

7.2.2 Municipal Solid Waste Management in Low- and Medium-Income Countries

To date, municipal solid waste management (MSWM) systems are at a critical stage in almost all the developing countries across the globe. Uncontrolled increases in population, weak institutional capacities, inadequate legislation, nonavailability of funds for infrastructure, and the poor behavioral patterns of the public toward waste management are some of the reasons behind the failure of the main technologies in SWM in various developing countries. In developed nations, the focus has now been shifted from waste- and sanitation-related diseases to "diseases of affluence" such as cancer, obesity control, heart-related illnesses, and sustainable development (Konteh 2009), whereas most developing nations are facing both problems simultaneously—that is, diseases of affluence and communicable diseases. Local governments are unable to deal adequately with these problems together. Wilson (2007) states that most developing countries are deprived of basic survival and infrastructure facilities, such that the local governments and the public are not much concerned about SWM facilities. In the near future, whenever SWM becomes a public agenda in developing countries, it will be based on the same drivers as in developed countries, although the current context is very different from that of developed nations in the past (Wilson 2007; Coffey and Coad 2010). The key priority for most governments is to improve public health and to remove solid waste from underfoot. People in developing countries consider waste a resource; the selling of recyclables provides them with a livelihood. In countries such as India, informal recyclers play a significant role in the collection of recyclables from solid waste; as a result, a huge amount of original material is saved from disposal (UN-Habitat 2010). Rapid urbanization, economic imbalances, weak institutional capacities, international influences, and varying socioeconomic and political landscapes together have produced highly complex challenges for governments in developing countries. The main challenges that SWM systems are facing in developing countries are mentioned in the following sections.

7.2.2.1 Urbanization

Urbanization is increasing at a rapid pace all around the world: Approximately more than 50% of the population has shifted to urban areas. This has put immense pressure on local governments to fulfill the demands of an increasing population. At the start of the twenty-first century, more than 400 cities around the world had populations of more than one million; in fact, three-quarters of these cities are located in developing countries (Cohen 2004). All the conventional infrastructural facilities present in developing countries are inadequate to serve the demands of ordinary people. The local governments of the burgeoning cities in developing countries are unable to provide quality SWM and sanitation services. Most of the populations in towns are living in densely packed slums where there is no space to dump, collect, or decompose waste, and people cannot make any alternative arrangements to dispose of their waste. Slum areas are so unplanned that it is impossible to find even small spaces to provide waste containers, and narrow and undulating roads hinder the path of waste collection vehicles (Coffey and Coad 2010). As a result, solid waste is openly dumped onto roads and into waterways, which become clogged in rainy seasons and are the birthplace of many communicable diseases (Coffey and Coad 2010).

7.2.2.2 Cultural and Socioeconomic Aspects

Any waste management system is highly influenced by the behavior and attitude of the local population of a region. Every component of a waste management system, such as daily waste generation, storage in containers, reducing consumption, reusing, recycling, willingness to pay for SWM services, and reactions toward proposed SWM facilities, is affected by the level of public awareness and by public attitudes. In Middle Eastern countries, working as a solid waste professional is considered a low-grade profession; this results in weak waste management institutions (Wilson 2007).

Socioeconomic status and cultural values vary between regions, and this influences the waste composition; high literacy levels are equivalent to high paper and plastic contents in solid waste. High-income populations generally discard waste that can be either reused, recycled, or repaired; this ultimately increases their daily waste generation and consumption rate. In most developing countries, basic needs such as food, water, shelter, and security are people's main priorities; SWM will only become a priority either when these basic needs have been met or when they are affected by poor SWM (Konteh 2009).

7.2.2.3 Political Landscape

Politics plays a vital role in the implementation and operation of SWM schemes in any country. A major challenge for many developing countries is to create a proper balance between policy, governance, and institutions.

1. *Policy*: In most developing nations, weak policy measures and their improper implementation are the main causes behind the failure of SWM systems. Mostly policies are not framed as per the needs and demands of citizens; as a result, governments are unable to allocate proper resources for the functioning of SWM systems, whereas developed nations formulate proper policies, considering the present and future demands of their citizens. Strong institutions also help in the proper implementation of policies.

2. *Governance*: Any form of environmental management in any country around the world is a critical political task; different governments have different interests and aims, such as competing for the ownership of land, for resources, and for infrastructure and services (Hardoy et al. 2001). If environmental management is absent from the public agenda, it will result in resource depletion and environmental health degradation. The main task of government is to focus on the roles, responsibilities, and functions of its bodies, but this neglects the relationship between government and civil society. The concept of governance focuses on the participation and collaboration of the public, the government, nongovernmental organizations, and the private sector. Good governance practices help in the proper implementation of policies and the allocation of resources, which are essential for the functioning of SWM systems.

3. *Institutions*: According to Wilson (2007), weak institutions have become a major issue in many developing countries all around the world. As a result, the capacity building and strengthening of institutions has become a main driver for SWM. To run an effective SWM system, it is necessary to clearly distribute the roles and

responsibilities of government bodies, which will help in avoiding controversies and ineffective SWM systems.

7.2.2.4 International Influences

It has been observed that most SWM systems and projects in developing countries fail due to weak institutions and an absence of trained professionals in the SWM sector. Here, international financial institutions (IFIs) play a major role in capacity building and project funding. The main focus areas of IFIs in developing nations are removing poverty, building infrastructure, environmental protection, and capacity building (Marshall and Farahbakhsh 2013). Most of the time, the approaches adopted by IFIs in developing nations are not planned according to local contexts, due to poor understanding of the region and a lack of funds for the operation of facilities, as IFIs provide funding for infrastructural development but ignore the fact that developing countries may not be able to afford operational costs. According to Coffey and Coad (2010), the technology and machinery supplied by IFIs (whose focus is to capture local markets) are not completely appropriate for local regions, hence sometimes resulting in the closure of facilities after a few years of operation.

7.3 Need for Systems Approaches to Solid Waste Management

MSWM in any urban area is a most difficult and complex task, due to the number of streams, such as collection, transportation, treatment, disposal, and so on. All these factors are greatly influenced by public behavior, the local government's priorities and its capacity to handle waste, and the functioning of authorized companies for collection, transportation, and treatment. According to Seadon (2010), it is necessary to understand the interaction between the physical components, such as production, transportation, urbanization, and land use patterns, and conceptual components, such as the socioeconomic and environmental aspects, for the proper management of waste. Most developing countries follow the end-of-pipe concept, which means treating and managing mixed waste when it reaches the facility. A lot of problems are associated with mixed waste, which results in the failure of major technologies. The conventional SWM approach is reductionist in nature, in which each waste component and stream is broken down into parts and each part is considered separately; in this way, observers analyze the whole system but neglect the relationship between various waste streams (Seadon 2010). This type of approach generally provides piecemeal solutions that are not sustainable and are vulnerable to failure in the long run. There is a need to develop a systems approach that considers all aspects—social, environmental, economic, political, and cultural—during the planning of any sustainable SWM project.

7.4 Integrated Solid Waste Management: Systems Perspective

According to the systems perspective, ISWM focuses on all the waste streams—collection, disposal, and treatment—as subsystems and relates them in an integrated manner. To achieve sustainable waste management, a system should be designed in such a way that it will be environmentally effective, socially acceptable, and economically affordable (McDougall et al. 2001). An ISWM system should be integrated in such a way that it includes all waste material, is market oriented and resists fluctuations in market conditions, and is flexible enough for the inclusion of changes in consumer behavior over time (McDougall et al. 2001). In the present scenario there is no best ISWM system available; it is still at the conceptual level and refers to designing and implementing new systems and optimizing the existing ones. An ISWM system is a host of various systems in which every system functions differently, and because of higher-level complexity it is difficult to focus on every system, thus resulting in a trade-off at all levels. To implement ISWM systems and to manage waste systematically, there is an urgent need for strong regulations in both developed as well as developing countries, including programs focused on "closing the loop"—that is, shifting policy from end-of-pipe to resource management (Wilson 2007). Many developed and developing countries do not have actual or real ISWM systems because of barriers to policy implementations and a lack of systems thinking in their design. The next section focuses on different types of systems approach, and elaborates on systems engineering principles and types of system thinking.

7.5 Systems Approaches

Figure 7.2 shows various approaches for SWM. Disciplinarity and multidisciplinarity are two approaches that are based on reductionist and cause-and-effect concepts. Reductionist concepts break the whole of waste management into smaller parts and then deal with each part separately, whereas cause-and-effect thinking splits everything into parts and considers the relationship between parts but does not consider unmeasured variables and new relationships that may occur unexpectedly. The pluridisciplinary approach considers the cooperation between all the waste streams but lacks

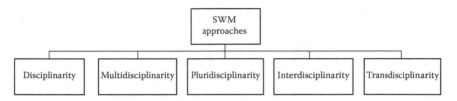

FIGURE 7.2
Waste management approaches.

coordination, whereas the interdisciplinary approach has a higher level of coordination. The transdisciplinary approach is the best approach to dealing with solid waste as it considers SWM systems as a whole. In this approach, SWM is considered a system in which not only every element and their internal relationship is studied but there is a focus on the generation of dynamic processes and the emergence of new properties at all levels (Seadon 2010).

7.6 Systems Engineering Principles

7.6.1 Systems Definition

A system is a set of elements that are interconnected with each other via certain relationships and which all function together to achieve a common goal (Meadows 2008). In this way, a system is greater than the sum of its parts. Every element and subsystem inside a system is defined by its specific functions and different roles at different levels. Complex systems theory explains that every wider system consists of several subsystems and elements with their own set of features or states inside a defined boundary. A wider system is comprised of various subsystems, where each subsystem has different elements with varying features; these all function together to form a unique structural relationship that implies the transfer of energy, material, and information (Chang and Pires 2015). According to I.D. White (1984), every system possesses some common characteristics, such as its structure, which depict real-world problems. Structural and functional relationships exist among all the units of a system. White also states that a driving force or source of energy is required to transfer or interchange energy, material, and information across system boundaries.

To understand wide complex systems, boundaries are drawn that simplify the larger systems. System boundaries, by which systems are separated from the rest of the world, are imaginary in nature. According to Meadows, there is no rule to define a system boundary; it is drawn according to the needs of analysts and researchers for the understanding of wide systems. The behavior of a system boundary assists in understanding different types of systems. On the basis of its boundary, a system is defined as isolated, closed, or open. An isolated system never interacts with its surroundings; such systems are observed only at laboratory scale. In closed systems, material never transfers across boundaries, only energy. Open systems are those where energy, material, and information flow across the boundary and are shared between elements and subsystems. As system boundaries are imaginary in nature, the main problem arises when they are drawn either too narrow or too broad; narrow boundaries always create complicated problems, whereas too broad a boundary makes a system more complex to understand. A perfectly defined system boundary should be based on a specific context and helps in understanding what is included in analysis and what is not (Meadows 2008).

7.6.2 Systems Thinking

The concept of *systems thinking* was borne out of von Bertalanffy's mathematical model of the *general systems theory* (GST) in 1937. The main aim behind GST was to unify science by applying the principles of living systems on all systems (Seadon 2010). GST was unable to go beyond the general concept of holism, but it has sparked the concept of the systemic approach among scientists and researchers. Systems thinking was an attempt to shift away from the concept of reductionism of normal science to a transdisciplinary approach. Systems thinking focuses on the relationship between the various elements of a system, along with their patterns of behavior, processes, and contexts; this result in a better understanding of the interdependency between a system and its subsystems. Systems thinking was developed from the hard science of mathematics; scientists found it inefficient in dealing with real-world issues, and this led to the emergence of *soft systems thinking*. *Hard systems thinking* follows a well-defined approach toward solving real-world problems by defining a problem then analyzing the situation to identify objectives that help in measuring performance and developing options to implement change. On the other hand, soft systems thinking uses systems as intellectual devices that derive from the views of the relevant stakeholders to understand a problematic situation and find solutions for it (Checkland 2000). Therefore, hard systems thinking can be used for well-defined, technical problems, whereas soft systems thinking is used to find solutions for messy and poorly defined situations that involve social and cultural aspects (Marshall and Farahbakhsh 2013; Checkland 2000).

7.6.3 Systems Engineering Approaches

Systems engineering is an interdisciplinary field whose main aim is to develop complex organized systems in which each element or part has a defined role to generate a proposed result. The major goal of systems developed by using systems engineering principles is to provide quality products or solutions that ultimately fulfill the needs of consumers. Systems engineering should involve technical, social, economic, and managerial factors in developing solutions and technologies—independent techniques that can handle complex real-world situations (Chang and Pires 2015). Solutions developed from systems engineering principles are such that they will help in decision-making during the planning process. While proposing alternative solutions during the planning process, systems engineers should include all the technical, financial, legal, and managerial constraints and the perspectives of all the stakeholders (Chang and Pires 2015). Top-down and bottom-up design are two alternative approaches available to systems engineers.

Generally, the top-down approach breaks the system into subsystems to gain a better understanding of the relationship between different subsystems at different levels. Further, the subsystems are broken down to base elements. In this approach, an environmental problem is first defined, then system boundaries are drawn and the objectives of the decision-makers are identified accordingly. Then, alternative solutions are produced based on simulation, optimization, or forecasting techniques (Chang and Pires 2015). After the evaluation and screening of alternative solutions, the most appropriate and feasible solution is implemented to check the performance of the system.

The bottom-up approach is used to find the limits of an engineered system; this is done with the help of various mathematical tools such as differential equations and partial differential equations that simplify the complex system. In this approach, the base elements of systems are mainly focused on in detail. To understand these elements, a general perception is formed on the basis of data collected from different stakeholders or external environments. Then, these elements are linked together to form subsystems that are further linked at a higher level to form a complex system. In this way, the bottom-up approach is used to engineer a complete system in order to understand and handle the complexity of the situation.

7.7 System-of-Systems Approach

During the development of systems theory, a new approach was developed to understand the problems of large scales, which is defined as the system-of-systems (SoS) engineering approach. This approach brings the various independent systems together to develop a more complex system of greater functionality to handle problems at large. SoS does not provide any new tools or methodology to deal with complex problems, but it provides a better thinking approach to finding solutions based on the interaction of technology, policy, and government. Maier (1998) explained five characteristics of the SoS approach: the geographic distribution of elements, the independent operational and managerial elements, the emergence of new properties and behavior, and their evolutionary pathways. SoS studies in depth external as well as internal factors that affect the planning, designing, and operational phases of waste management (Chang 2011). The major external factors covered under SoS are urbanization, resource exploitation, climate change, and political influences, whereas internal factors include waste composition, treatment, and regional variation (Chang 2011). Using SoS, a proper understanding of the external and internal factors that affect ISWM in developing countries assists in finding the advantages and disadvantages associated with centralized and decentralized waste management systems. The next section focuses on centralized and decentralized waste management systems for developing countries.

7.8 Centralized and Decentralized Systems

The selection of centralized and decentralized systems to deal with SWM challenges depends on the willingness to regionalize; moreover, the success of regionalization plans depends on the willingness of local organizations and their heads. A centralized system for SWM should include economic, environmental, and social aspects such as life cycle studies of different parts of SWM and incentives for sustainable SWM practices. Waste hierarchy principles should be followed, treatment and disposal facilities should be built near human settlements, and policies should be developed to address the risk associated with human health. It has been observed that most decentralized systems are designed based on the experience gained from local SWM, and these experiences help in initiating local businesses that promote decentralization (Chang and Pires 2015). Centralized waste management systems represent complex and large-scale capacities with low operational costs and are managed by municipal corporations or urban local bodies, whereas decentralized systems are locally owned and operated by residents or authorities of a small region (Chang and Pires 2015). Centralized systems have a higher risk of affecting environmental and human health as they handle large amount of waste; in contrast, decentralized systems have less risk because a small amount of waste is distributed among various communities at different locations and treatment facilities can be constructed in close proximity to human settlements. Generally, decentralized approaches are practiced in low- and medium-income countries because of the low initial investment and operational costs, whereas centralized approaches are practiced in high-income countries that can afford high initial investment for the establishment of SWM facilities (Medina 2010).

7.9 Systems Analysis Techniques for Municipal Solid Waste Management

System analysis is a technique to solve [a] problem. It analyzes each and every segment of the component in order to scrutinize the whole data so that the system should work in a smooth manner. Nowadays, a new approach to system dynamics modeling is being developed that can predict SWM generation amid rapid urbanization and a limited sample (Chang et al. 2005). And in the last few decades, a market has emerged that uses systems analysis to handle MSWM through a range of interrogative methodologies, so as to remedy the challenges, tendencies, and perceptions of legacy analysis systems. The main benefit of using systems analysis is that it assists in developing long- and short-term MSWM strategies that involve social, economic, and environmental aspects (Baetz 1990; Thomas et al. 1990). In the present scenario, to develop SWM systems from sustainability and economic perspectives, it is necessary to formulate systems engineering models or systems assessment tools (Chang et al. 2011). A total of five systems engineering models and nine systems assessment tools were formally classified (Chang et al. 2015). It is worth noting that the spectrum of these models and assessment tools was classified based on the following two domains, although some of them may be intertwined with each other (Chang et al. 2015).

7.9.1 Systems Analysis Techniques

In systems engineering techniques, a system is correlated with subsystems. The characteristics and relationship of the system give information regarding the properties of the system (Pires et al. 2011). MSWM related to this theory gives information about its technical aspects, such as landfill, incineration, anaerobic digestion, or composting systems (Pires et al. 2011). Considering MSWM systems, several systems analysis techniques have been applied to help decision-making. These can be divided into two main groups:

- *Systems engineering models*: Systems engineering models promote analysis based on cost–benefit analysis, optimization models, simulation models, forecasting models, and integrated modeling systems (Pires et al. 2011).
- *Systems assessment tools*: Various systems assessment tools such as scenario development, material flow analysis, life cycle assessment or life cycle inventory, risk assessment, environmental impact assessment, strategic environmental assessment, socioeconomic assessment, and sustainable assessment can be helpful in analyzing the loopholes and providing the ultimate goal of MSWM by improving the system (Chang 2011).

A deep understanding of the development drivers and present challenges to SWM systems in developing countries will assist in formulating better solutions for persistent problems. To develop an ISWM system for developing countries requires major changes in existing systems. Various subsystems in ISWM systems are interrelated with each other; thus, making small piecemeal changes in existing systems will not help in achieving the total quality objective (achieving both environmental and economic sustainability). Therefore, it is necessary to conceptually redesign whole new systems (McDougall et al. 2001). Various systems analysis and engineering approaches can be applied to analyze the relations between various subsystems and the flow of information, material, and energy across system boundaries. A hybrid of hard and soft system approaches that considers all the aspects, technical as well as socioeconomic, can be used to develop a better ISWM system for developing countries.

8

Municipal Solid Waste
Management Planning

8.1 Introduction

To manage the enormous size of a municipal solid waste management (MSWM) system, proper planning is required. Planning is the first step in designing or improving a waste management system, taking into consideration environmental, social, economic, and technical factors. These are essential to setting achievable goals. As factors differ from one place to another, it becomes a challenge to handle waste management issues at different locations.

8.2 Effects of Improper Planning for Implementation of MSWM Systems

Municipal solid waste (MSW) dumps are seriously spoiling the environmental conditions in developing countries. Negative environmental impacts from the improper management of MSW dumping can easily be observed everywhere in the developing world. Inappropriate waste management has influential effects on the climate, on humans, and on land resources. It deteriorates air and water quality and also affects the land quality by encroachment. This extra burden causes global warming and climate change, thus affecting the entire planet. Improper management of waste subsequently causes many premature deaths (WTERT 2012).

Solid waste disposed of recklessly in the open sometimes ends its journey in the drainage system, resulting in the overflow of water, causing discomfort to people. In 2006, Mumbai faced a flood, one of the major reasons of which was the clogging of the sewage system. Disease outbreak was another dreadful result of the improper dumping of waste, as it attracted insects and rodents, spreading diseases such as cholera, dengue fever, and plague.

Polluted water is also unsuitable for bathing or irrigation, as it is too infected with diseases and other contaminants (Annepu 2012).

The open burning of MSW either in landfills or on streets releases various pollutants, such as carbon monoxide (CO), carcinogenic hydrocarbons (HC) including dioxins and furans, particulate matter (PM), nitrogen oxides (NO_x), and sulfur dioxide (SO_2). These pollutants degrade the quality of air. Also, solid waste dumped in landfills or on streets for a long time liberates greenhouse gases, ultimately participating in the greenhouse effect (WTERT 2012).

Inappropriate solid waste management in India, specifically, contributes to 6% methane (CH_4) emission; thus, this sector is the third largest emitter of methane. This amount of methane emission is much higher than the global average of 3% methane emissions from solid waste. Methane has 21 times more effective global warming potential than carbon dioxide (CO_2). MSW currently liberates 16 million tons of carbon dioxide equivalents per year and this number is expected to rise to 20 million tons by 2020 (Pawale et al. 2015).

Waste can also be used as a resource. Due to ignorance and a lack of knowledge about MSWM, it is estimated that developing countries such as India will lose 6.7 million tons of secondary raw materials or recyclables, 9.6 million tons of organic fertilizers, and 57.2 million barrels of oil (Annepu 2012).

Looking at the current scenario of waste management, it can be seen that there is an urgent need for appropriate MSWM in developing countries. A guidance note titled "Municipal Solid Waste Management on a Regional Basis" (2011) by the Ministry of Urban Development (MOUD), Government of India (GOI), observes that compliance with the "Municipal Solid Waste (Management and Handling) Rules" (2000) requires appropriate systems and infrastructure to be put in place to undertake the scientific collection, management, processing, and disposal of MSW. However, not much effort has been made in this respect and the authorities concerned are unable to implement independent schemes or plans for the scientific management of MSW, from the time of its collection to its disposal. This sector of waste management lacks financial and technical expertise, and has a scarcity of funds and manpower (WTERT 2012).

8.3 Requirements in MSW Planning

For the effective planning of MSWM, this sector needs to consider the entire life cycle of waste and the complete set of environmental effects and costs associated with the management practice. Worldwide federations of national standard bodies such as the International Organization for Standardization (ISO) have regulated the framework on life cycle assessment (LCA). But the regulations are not very effective in developing countries due to a lack of awareness of the importance of using the life cycle concept. These countries also lack a general understanding of how to conduct LCAs correctly and deduce the results. Emery et al. (2007) assessed the environmental impacts of a number of waste management scenarios; during the assessment an interactive model was created to judge the financial aspects, management aspects, and recovery rates using various waste recovery methods including curbside recycling and incineration. The LCA report stated that treatment options such as incineration came out as better ideas than landfill, recycling, and composting. The study also indicated that *integrated municipal solid waste management* (IMSWM) is the most appropriate methodology in terms of both the economic and environmental benefits (Ramachandra 2011).

8.3.1 Factors Affecting IMSWM

Waste management hierarchy in simpler terms incorporates IMSWM (Turner and Powell 1991; Tchobanoglous et al. 1993). The waste management hierarchy shown in Figure 8.1 is influenced both by direct and indirect factors.

These factors lay a framework that can be used to optimize the existing system as well as help to design and apply a new waste management system (UNEP 1996; 2002). IMSWM also handles media such as solids, liquids, or gases. Pimenteira et al. in 2005 analyzed solid waste management systems both socially and economically, and found that reusing secondary products of the recycling process can reduce greenhouse gas emissions (Ramachandra 2006).

8.3.2 Planning IMSWM Systems

Formulating an IMSWM needs a concrete approach; that is, it should deal with all types of waste along with the management procedure, right from generation to disposal. IMSWM for any city should ensure public health and

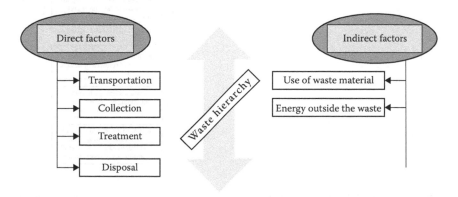

FIGURE 8.1
Direct and indirect factors of IMSWM.

safety. In addition, an effective IMSWM must be environmentally and economically feasible. There should be balance and control within the system, and it should be capable of reducing the overall environmental impacts of waste management within an acceptable level of cost (Ramachandra 2011).

An integrated approach must be based on a logical hierarchy of actions. The main goals of IMSWM are shown in Figure 8.2.

For optimal solid waste management, a *geographic information system* (GIS), a spatial tool, can be used to formulate a framework for waste management. This tool can be helpful in finding a source segregation and collection system. With the help of this tool, the production of biogas, hazardous waste management, and safe disposal can be figured out (Ramachandra 2006).

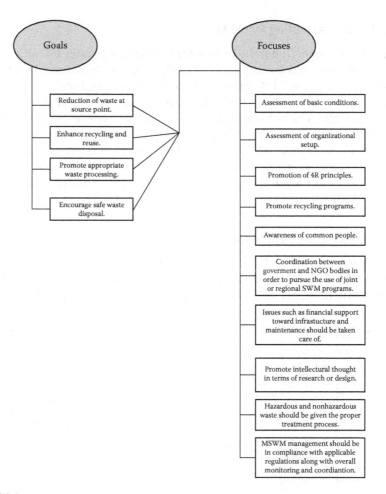

FIGURE 8.2
Goals and focus of IMSWM. (Modified from Ramachandra, Integrated management of municipal solid waste. *Environmental Security: Human and Animal Health*, pp. 465–484. Lucknow, India: IBDC, 2011.)

8.4 Tactical and Strategic Planning for Implementation of MSWM Systems

Tactical and strategic planning for the implementation of MSWM systems must be maintained and followed through a wide variety of functional networks. The overall function should follow the IMSWM system in the following ways:

- Life cycle–based IMSWM

    The initial stage of IMSWM begins with an LCA of a product, keeping in mind its production and consumption patterns. If the consumption pattern is reduced and utilization of the discarded product is increased, then the burden on the primary resources will also be reduced, thus implying less effort on the final disposal of waste. Figure 8.3 represents life cycle–based IMSWM.

- Generation-based IMSWM

    The second stage of IMSWM is to categorize the waste emerging from different sources, such as household waste, commercial waste, and agricultural waste. It could be further categorized according to its hazardous and nonhazardous nature. Hazardous waste should be separated and isolated from the rest, and proper disposal techniques should be used according to regulations. The *4R approach* (reduce, reuse, recycle, and recover) is a must for the management of waste, right from its collection to its disposal (UNEP 2009). Generation-based IMSWM is shown in Figure 8.4.

- Management-based IMSWM

    The basis of the third stage of IMSWM is management; this includes the following, according to UNEP (2009):

- Laws and regulations
- Mechanisms to increase economy
- Technology to deal with the treatment of waste
- Participation of stakeholders in the waste management system

8.4.1 Strategic Planning of MSWM Systems

To develop the strategic planning of MSWM, different issues are taken into consideration. First, a baseline is defined to establish a strategic planning framework, and then there is an evaluation and identification of the available options. Strategies are taken into consideration so that the plan can be effective and sustainable for a longer period.

The process identifies six concrete steps, shown in Figure 8.5, toward strategically planning MSWM practices.

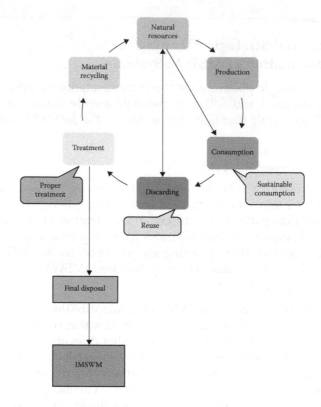

FIGURE 8.3
Life cycle–based IMSWM. (Data from UNEP, *Developing Integrated Solid Waste Management Plan Training Manual*, Vol. 2. Osaka, Japan. Assessment of Current Waste Management System and Gaps Therein. United Nations Environmental Programme Division of Technology, Industry and Economics International Environmental Technology Centre 2009.)

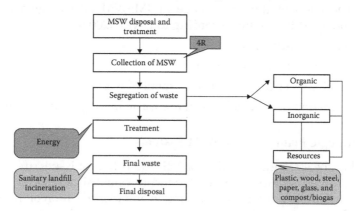

FIGURE 8.4
Generation-based IMSWM. (Data from UNEP, *Developing Integrated Solid Waste Management Plan Training Manual*, 4, 66–88. Osaka, Japan. UNEP, Division of Technology, Industry and Economics International Environmental Technology Centre, 2012.)

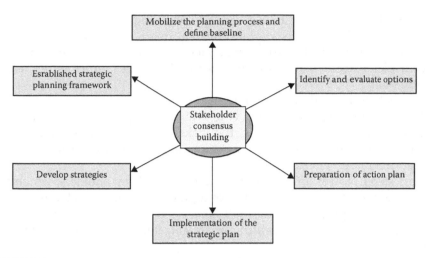

FIGURE 8.5
Steps toward strategically planning municipal solid waste management practices. (Modified from VNG International, *Municipal Development Strategy Process: A Toolkit for Practitioners*, The Hague, The Netherlands, VNG International, 2010).

Each of the steps identified includes subactivities and has a list of expected outcomes.

Step 1: Organize the planning process

- The first step of strategic planning is a memorandum signed by the key stakeholder in order to show their commitment to the plan. It is a type of agreement or pact known as a *memorandum of understanding* (MoU). It legally binds the stakeholder to his or her service.

- A steering committee is formed in order to direct and scrutinize the strategic planning process. The steering committee is composed of people who have expertise in technical or other aspects in order to run the strategic plan.

- The third step of organizing a strategic plan includes drawing up the *terms of reference,* a document that establishes the responsibilities carried out by the working people of the steering committee. Periodic scrutiny of the output at each step should also be ensured. It also includes the financial aspects of the project needed to raise the budget.

Step 2: Identification of the baseline
In this step, a basic study is done in order to get a glimpse of the existing MSWM system. The key issues relating to the MSWM system are required to protect the planning from any loopholes.

Step 3: Framework to establish strategic planning

A framework in strategic planning requires a plan for the long term, usually 10–20 years. It also incorporates assets in waste management. A 5-year-plan is framed at this stage. The first 1–2 years require an immediate action plan.

Step 4: Identify and evaluate options

The main strategy of this step is to evaluate the possible solutions to the problem. Preference is given to the stakeholder's preferred option. This option is then carried out at the beginning of the final strategy step.

Step 5: Strategy development

This step involves a final strategy agreed by all stakeholders, which acts as a guideline for the preparation of a plan of action. It includes a list of draft strategy proposals or a *draft strategy*. It also includes intermediate strategies, such as a list consisting of draft strategy proposals and the draft strategy.

Step 6: Preparation of the action plan

Keeping in mind the priorities, objectives, and targets, an action plan is framed for the first 2–5 years.

Step 7: Implementation of the strategic plan

For a successful MSWM system, implementation of the strategic plan is a must. The plan should be well supported by the stakeholder (Wilson et al., 2001).

8.5 Long- and Short-Term Planning for MSWM Systems

Planning is the systematic process of carrying out the intended task. Planning acts as a guideline to specify the needs, time, and priorities of the action concerned.

The planning of the intended task includes

- Recognition of the intended problem
- Collection of data
- Assessment of the problem
- Suggestions for intended actions
- A framework to formulate the strategy in order to solve the problem
- Evaluation of the action taken to combat the situation, along with modification of the plan if needed

A plan for a city should be documented in writing. It should outline the roles and responsibilities of the civic body, along with a set of principles and directives for achieving the goal of the project in a given time frame.

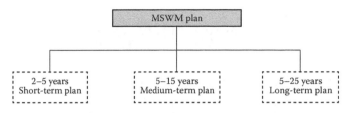

FIGURE 8.6
Different plans for MSWM.

An MSWM system is composed of various activities, such as the generation, storage, collection, transport, and disposal of solid waste. These different steps also incorporate aesthetic, economic, legal, and technical aspects, which altogether form an interdisciplinary relationship.

During the preparation of an MSWM plan, a particular time frame should be taken into consideration. Depending on the duration of the plan, an MSWM plan is divided into short-, medium-, and long-term plans. Figure 8.6 shows an MSWM plan and its duration.

A functional plan needs support from the local, state, and national levels so as to coordinate with the priority and available resources. The management practice should invariably be given strict guidelines regarding concerns of the generation stage to the disposal stage. MSWM needs both planning and design with accurate knowledge of the quantities of waste generation. Exact information helps any system to run smoothly. MSW generation in developing countries is quite challenging, as the countries are facing rapid urbanization and fast growth. Developing countries lack historic records of waste quantities and management procedures. Lack of information on this front greatly affects long- and short-term plans (Chang et al. 2005).

Strategies for MSWM are not up to the mark due to inappropriate technical and design aspects. The agencies responsible for waste management pay less attention to IMSWM and the interactions between systems and subsystems.

In order to improve an MSWM management system, proper attention should be given to the following aspects:

- Formulation of a plan, either strategic or tactical
- Cost–benefit analysis
- Budget of the intended project
- Calculations of cost per unit
- Economic analysis
- Environmental concerns
- Appropriate management options

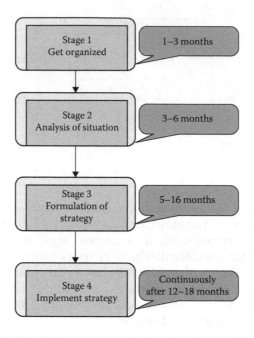

FIGURE 8.7
Strategies regarding long- and short-term planning. (Modified from VNG International, *Municipal Development Strategy Process: A Toolkit for Practitioners,* The Hague, the Netherlands: VNG International, 2010.)

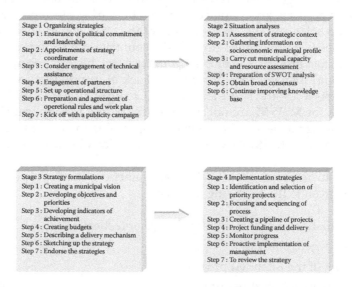

FIGURE 8.8
Substages of strategies for long- and short-term planning. (From WastePortal. net, Municipal strategic planning for solid waste management: Concepts and tools. http://wasteportal.net/en/waste-aspects/environmental-and-health-aspects/municipal-strategic-planning-solid-waste-management-c).

Other aspects that are important for effective MSWM are as follows.

- Techniques required for the collection of data
- Analysis of waste composition
- Waste generation projection techniques
- Technical expertise
- Procurement procedures
- Information regarding management systems
- Evaluation of the plan at regular intervals

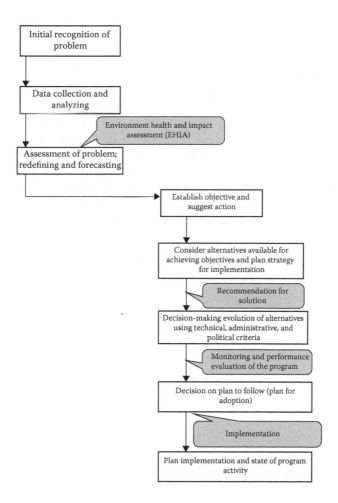

FIGURE 8.9
A basic planning model. (From The Central Public Health and Environmental Engineering Organisation (CPHEEO). *Manual on Water Supply and Treatment.* 2000. http://cpheeo.nic.in/Watersupply.htm.)

Thus, for managing and implementing MSWM systems, complete long- and short-term plans for their implementation need to be built up, depending on the specific requirements of the conditions. That is why systems analysis techniques for MSWM are required; only then can a complete solution for an MSWM system be worked out. Some strategies regarding long- and short-term planning are given in Figure 8.7, which shows four concrete stages for the creation of a municipal strategy along with their corresponding duration.

Each stage consists of a number of substeps, shown in Figure 8.8, which define the type of outcome expected.

So whether to go for long- or short-term planning, we must understand the basic planning model, as it helps to know which option is better to choose for a particular MSWM.

8.6 Basic Planning Model

The strategy of planning may appear stationary on paper, but in reality it is an active and continuous process. Each phase in a documented plan is considered a section in a civic body plan. A planned model provides feedback as it undergoes various events. The model, through its feedback mechanism, allows control of a correlation of errors. The model of a plan is also influenced by external factors, and this may cause sudden reversion in the model. Figure 8.9 shows the basics of the planning model.

9

Models for Municipal Solid
Waste Management Systems

9.1 Models for Community Bin Collection Systems

Any model cannot be adopted blindly. All models need to be validated as per the local conditions and as per the requirements. One must remember that every model has been designed as per the requirements of different projects and adopted according to the basic scenario. So any model must be evaluated and validated prior to being used for any predicted activity.

Solid waste is generated from different sections of society, such as households, corporations, markets, commercial areas, institutes, religious places, and so on. Mostly, in developing nations, unsegregated waste is collected and disposed of in open dump sites. The different types of waste generated and treatment practices followed in developing nations are shown in Figure 9.1.

Mostly, in developing nations such as India, waste is unsegregated, but if it can be separated according to the types of waste component, then it can be treated separately. According to its characteristics, municipal solid waste (MSW) is mainly composed of biodegradable, nonbiodegradable, and inert waste. The biodegradable fraction is further composed of organic waste, such as kitchen waste, fruit waste, garden waste, wood chips, and so on, while nonbiodegradable waste is composed of plastics, paper, Thermocol, cardboard, glass, leather, rubber, sanitary waste, metals, and so on. Of these wastes, plastics, paper, rubber, and metals are recyclable. Finally, the inert component is composed of sand, stone, bricks, and so on. Sometimes, household biomedical waste is also mixed with MSW (Boss et al. 2013). Table 9.1 highlights the waste compositions from different continents, and it shows that very high quantities of biodegradable waste are being generated from the Asian and African continents. Hence, they must be segregated and treated separately. If waste is segregated, then its commercial value can be obtained via proper processing and usage.

In developing nations, waste (largely unsegregated) is mostly disposed of in single community bins. Figure 9.2a shows the disposal of waste near

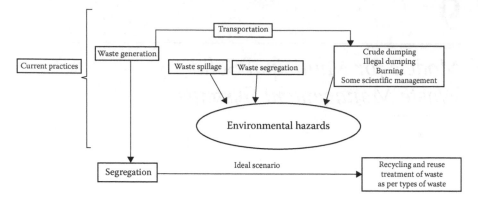

FIGURE 9.1
Current practices and ideal waste management practices for developing nations.

TABLE 9.1

Waste Composition of Different Continents

Continent	Organic (%)	Paper (%)	Plastic (%)	Glass (%)	Metal (%)	Inert (%)
Europe	35	21	10	6	4	23
Asia	51	14	11	4	4	16
South America	51	15	11	3	3	18
North America	42	22	13	3	6	14
Australia	52	18	8	5	4	13
Africa	56	9	10	3	3	20

Source: Modified from World Bank, Waste generation, Urban Development Series, Knowledge Papers, 2010. http://siteresources.worldbank.org/INTURBANDEVELOPMENT/Resources/336387-1334852610766/Chap3.pdf.

a dustbin. It has been observed that people mostly dispose of waste at the side of and not inside community bins. Boss et al. (2013) have reported that biomedical waste is being mixed with MSW, which affects the composition and handling mechanisms of MSW. Hence, they recommended segregating and treating biomedical waste separately. In some places, the illegal dumping of waste has also been observed on roads and in open spaces.

In order to properly segregate waste, one needs to carry out proper awareness and behavioral changes. Upadhay et al. (2012) have reported the achievement of waste segregation at the Malaviya National Institute of Technology campus in Jaipur, India. This was due to the appointment of dedicated staff and proper human resource development. In Germany, three different types of waste bin are placed for the collection of MSW: residual waste

FIGURE 9.2
(a) Waste disposed near dustbins. (b) Illegal dumping of waste.

bins, recyclables bins, and biowaste bins. People dispose of their waste in the respective bins, which are then transported by municipal bodies to the appropriate location for treatment (Schwarz-Herion et al. 2008).

9.2 Models for Vehicle Routing

Vehicle route planning is necessary for optimum usage in terms of waste collection, and proper planning will also aid in saving time, manpower, and fuel costs. If a properly planned routing system is in place, it will aid in the optimization of the collection system. Some of the models that are used for vehicle routing are discussed in Table 9.2.

Among these various models, Tavares et al. (2009) used geographic information system (GIS)-based three-dimensional (3-D) mapping for predicting waste collection routes. They also focused on reducing fuel demand while planning out a sustainable waste collection route in Cape Verde. They considered various factors such as routes, fuel consumption, and emission loads, and, based on those, an optimized route was designed. In order to design a road network model for MSW collection, various features must be taken into consideration, as given in Table 9.3.

Thus, in order to properly plan for waste collection, one needs to include the various parameters while chalking out routes, along with the best possible results. Based on this, an optimal route must be created that should have less time for collection services, fewer collection points, less fuel consumption, and less fuel demand. All these factors must be cross-validated and monitored. One must be open to changes and to revalidating the route once designed. Each time, one must be willing to redesign and rework the best possible options to achieve the desired result. The details of such a model are given in Figure 9.3.

TABLE 9.2

Vehicle Routing Models

Name	Description and Remarks	References
Dijkstra's algorithm	Used to plan the shortest network. Cannot be used for long city-based planning systems.	Dijkstra (1959)
A* search	A technique heavily used in artificial intelligence. Directs the search of Dijkstra's algorithm toward the target by using lower bounds on the distance to the target.	Goldberg and Harrelson (2005)
Reach-based routing	The *reach value* is defined as the minimum length of the subpaths dividing the shortest path at the node, and maximized over all the shortest paths. It is the shortest path to reach the target.	Sanders and Schultes (2006)
Highway hierarchies	Provides an alternate path to reach the destinations through the shortest route.	Sanders and Schultes (2006)
Highway node routing	Provides details of shortcuts to reach a particular destination.	Geisberger (2011)
Contraction hierarchies	Assigns a path depending on the importance level of the road and target.	Geisberger (2011)
Transit nodes	Highlights the shortest distance to reach the target. This model is used mainly for planning long-distance networks.	Bast et al. (2007)

9.3 Landfill Gas Modeling

A majority of the reactions taking place in waste degradation landfills are catalyzed by naturally occurring microorganisms. The common practice when measuring the performance of the solid waste degradation process is to measure the rate at which microorganisms metabolize the waste, which can be related to their rate of growth and gas production. Various models are available to predict landfill gas (LFG) emission, especially methane from open dumps or landfill sites (Sil et al. 2014). Some of the models are IPCC models (IPCC 1997; 2006), the Shell Canyon model (Thompson et al. 2009), and LandGEM (USEPA 2005). A comparison of the different models is presented in Table 9.4.

TABLE 9.3

Factors to be Considered for Route Planning

Factors	Description
Population density and waste generation	Waste generation is directly proportional to population density. The higher the density, the more waste generation. Hence, the population level and the rate of waste generation must be taken into account while designing any waste collection route.
Waste characteristics	Depending on the type of waste, the plan for a collection system can be chalked out. For instance, in developed countries, higher quantities of dry waste are generated compared with developing and underdeveloped nations. There are separate treatment methods for dry and wet waste, hence they must be collected separately for treatment. Waste collection points can be designed based on waste characterization.
Level of segregation	Again, waste must be segregated and disposed of, which is dependent on the behavioral patterns of the population. One must be encouraged to carry out proper segregation. Depending on the level of segregation, community bins or collection points can be designed.
Collection points	Collection points must be created depending on the population density and the level of waste generation.
Road width and capacity	Various types of vehicle, varying in size, collect and dispose of waste. If the road width is small, then smaller vehicles must be deployed for the collection of waste. Thereafter, waste from these smaller vehicles must be disposed of at secondary storage units, and from there it can be carried to treatment and processing plants.
Route length and optimization	Waste collection points must be designed in such a way that most of the vehicles need not travel more than 15 km for the collection of waste. Route maps need to be optimized for the level of waste generation, population density, and road width. If the roads are smaller in width, then hand-driven vehicles can be used. Again, the route length must be about 3–5 km. Then there must be another collection point that will carry the waste to disposal sites.
Route generation and time criteria	One must create time zone planning for waste collection. While designing waste collection routes, one must ensure an appropriate time for collection. Mostly, waste is collected in morning sessions. Hence, the shortest and most efficient path must be created for the waste collection mechanism. This will aid in efficient collection and fuel consumption.

Source: Adapted from Bhambulkar and Khedikar, *International Journal of Advanced Engineering Technology*, 2(4), 48–53, 2011; Mrówczyńska, *Transport Problems*, 9(1), 61–68, 2014; Tavares et al., *Waste Management*, 29, 1176–1185, 2009.

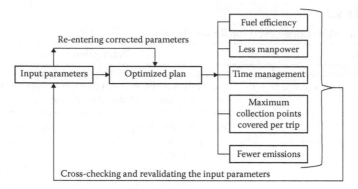

FIGURE 9.3
Plan for an ideal routing system for waste collection.

The US Environmental Protection Agency (USEPA) Landfill Gas Emissions Model (LandGEM) is one such model that is used extensively in India for predicting LFG emissions (Thompson et al. 2009). LFG models continue to receive criticism due to their poor accuracy and insufficient validation. The results of most of the models have not been evaluated against the methane recovery data (Barlaz et al. 2004). A few studies (Spokas et al. 2006; Barlaz et al. 2004) have compared the methane recovery data with estimated methane generation models, but only for a few landfills. Thompson et al. (2009) compared various models for methane emission from various landfill sites and concluded that the LandGEM model estimated methane emissions with better accuracy compared with other models. However, several other models are available for estimating the rate of LFG generation. A mass balance approach is the simplest emission estimation method, using the stoichiometric equation as described in Section 6.3.4. However, it generally overestimates emissions. Other useful models for LFG emissions estimation are the Scholl Canyon model, the Intergovernmental Panel on Climate Change (IPCC) First-Order Decay (FOD) model, the IPCC default method, and the triangular method (as described in Chapter 6) (Kumar et al. 2004; USEPA 2005; IPCC 2006; Thompson et al. 2009). These models vary widely, not only in the assumptions that they make, but also in their complexity and the amount of data they require. Sil et al. (2014) validated the LandGEM estimations with laboratory-scale biochemical methane potential (BMP) assays. It is important to validate the results of these models, as most of these results are used to predict the clean development mechanism (CDM) potential and financial benefit of the test sites (Unnikrishnan and Singh 2010). The findings of Sil et al. (2014) showed that a correction factor is needed to predict methane emission: 0.29 ± 0.13, 0.94 ± 0.02, and 0.54 ± 0.32 for mixed unsegregated waste, food waste, and vegetable waste, respectively.

TABLE 9.4

Comparison of Different LFG Models

Type	Order	Feature	Notes	References
EPER, Germany	Zero	Based on total waste and the proportion of biodegradable waste	Underpredicts the results; based on German conditions	Kamalan et al. (2011); Thompson et al. (2009)
SWANA	Zero	Based on weight of waste, time	—	Kamalan et al. (2011)
IPCC	Zero	Waste landfilled and degradable organic carbon	Depends on degradable fraction of waste; requires laboratory assessment	IPCC (1997; 2006)
TNO	First	Carbon content	Organic component not included	Amini et al. (2012); Scheutz et al. (2011)
GasSim	First, multiphase	—	No complete information	Donovan et al. (2010)
Afvalzorg	First, multiphase	—	Only applicable for the Netherlands waste composition	Kamalan et al. (2011)
EPER, France	First, multiphase	—	Complicated; requires a normalization factor	Amini et al. (2012)
Mexico model	First	Based on US and Mexico waste	Decay constant needed	Kamalan et al. (2011)
LFGGEN	First	Based on US waste composition	Methanogenesis preceding the lag phase	Barlaz et al. (2004)
Halvadakis	Complex mathematical model	—	Too hard to compute; only depends on carbon content	Kamalan et al. (2011)
LandGEM	First	US-based conditions	User-friendly and easily solved	Kamalan et al. (2011)

References

Agarwal, M. 2014. An investigation on the pyrolysis of municipal solid waste. In PhD thesis from the School of Applied Sciences, College of Science, Engineering and Health, RMIT University, Melbourne, Australia.

Agdag, O.M. 2008. Comparison of old and new municipal solid waste management systems in Denizli, Turkey. *Waste Management*, 29(1), 456–464.

Agyopng, J.S. 2011. Barriers in private sector participation in sustainable waste management experiences of private operators and waste service providers in Ghana. Zoomlion Ghana Limited Presentation at the UN on Building Partnership for Moving Towards Zero Waste.

Ahmad, K.N. and Mir, A.Q. 2014. Perspectives of transport and disposal of municipal solid waste in Srinagar city. *International Journal of Recent Research in Interdisciplinary Sciences (IJRRIS)*, 1, 7–16.

Ahmad, R., Jilani, G., Arshad, M. Zahir, Z.A., and Khalid, A. 2007. Bio-conversion of organic wastes for their recycling in agriculture: An overview of perspectives and prospects. *Annals of Microbiology*, 57(4), 471–479.

Ahmed, A.S. and Ali, M. 2004. Partnerships for solid waste management in developing countries: Linking theories to realities. *Habitat International*, 28(3), 467–479.

Ahn, H.K., Richard, T.L., and Choi, H.L. 2007. Mass and thermal balance during composting of a poultry manure: Wood shavings mixture at different aeration rates. *Journal of Process Biochemistry*, 42, 215–223.

Alberto, B. and Carlos, S.F. 2007. Waste management in developing countries: Present conditions and foreseen paths; A Brazilian overview (Report). São Paulo, Brazil: Brazilian Association of Urban Cleansing and Waste (ABRELPE).

Albiach, R., Canet, R., Pomares, F., and Ingelmo, F. 2000. Microbial biomass content and enzymatic activities after the application of organic amendments to a horticultural soil. *Bioresource Technology*, 75, 43–48.

Alburquerque, J.A., Gonzalvez, J., Tortosa, G., Baddi., G.A., and Cegarra, J. 2009. Evaluation of alperujo composting based on organic matter degradation, humification and compost quality. *Biodegradation*, 20, 257–270.

American Chemistry Council. 2015. Recycled plastic products can give homes a green makeover. https://www.plasticsmakeitpossible.com/plastics-recycling/what-happens-to-recycled-plastics/recycled-plastic-products-can-give-homes-a-green-makeover/ (accessed December 10, 2015).

American Society for Testing and Materials. 2013. ASTM D3172 standard practice for proximate analysis of coal and coke. http://www.astm.org/Standards/D3172.

Amini, H.R., Reinhart, D.R., and Mackie, K.R. 2012. Determination of first-order landfill gas modeling parameters and uncertainties. *Waste Management*, 32(2), 305–316.

Angelidaki, I., Karakashev, D., Batstone, D.J., Plugge, C.M., and Stams, A.J.M. 2011. Biomethanation and its potential. In A.C. Rosenzweig et al. (Eds), *Methods in Methane Metabolism: Methanogenesis*, p. 424. London: Academic Press.

Annepu, R.K. 2012. Sustainable solid waste management in India. Waste-to-energy research and technology council (WTERT) Bi-Annual conference, Columbia University. http://www.seas.columbia.edu/earth/wtert/sofos/Sustainable%20Solid%20Waste%20Management%20in%20India_Final.pdf.

Ansah, B., 2014. Characterization of municipal solid waste in three selected communities in the Tarkwa Township of Tarkwa-Nsuaem Municipality in Ghana, A thesis submitted to the Department of Environmental Science, Kwame Nkrumah University of Science and Technology, Kumasi, in partial fulfillment of the requirement for the award of Master degree in Environmental Science.

API. 2016. APINAT BIO bioplastics. http://www.apinatbio.com/eng/apinat-bioplastics.html (accessed May 26, 2016).

Asian Development Bank (ADB). 2013. Solid waste management in Nepal: Current status and policy recommendation. Mandaluyong, Philippines: ADB.

Asian Productivity Organization (APO). 2007. Solid waste management: issues and challenges in asia; report of the APO survey on solid waste management, 2004–05. Tokyo, Japan: APO.

Asokan, P., Saxena, M., and Asolekar, S.R. 2007. Jarosite characteristics and its utilisation potentials. *Science of the Total Environment* 2007, 359, 232–243.

Awasthi, M.K., Bundela, P.S., Pandey, A.K., Khan, J., Wong, J.W.C. and Selvam, A. 2014. Evaluation of thermophilic fungal consortium for organic municipal solid waste composting. *Bioresource Technology*, 168, 214–221.

Babayemi, J.O. and Dauda, K.T. 2009. Evaluation of solid waste generation, categories and disposal options in developing countries: A case study of Nigeria. *Journal of Applied Sciences and Environmental Management*, 13(3), 83–88.

Baetz, B. 1990. Optimization/simulation modelling for waste management capacity planning. *Journal of Urban Planning and Development*, 116(2), 59–79.

Baoyi, L.V., Xing, M., Yang, J., Qi, W., and Lu, Y. 2013. Chemical and spectroscopic characterization of water extractable organic matter during vermin-composting of cattle dung. *Bioresource Technology*, 132, 320–326.

Barber, R.D. 2007. Methanogenesis: Ecology. In *Encyclopedia of Life Sciences*, pp. 300–310. New York: Wiley.

Barlaz, M.A., Green, G., Chanton, J.P., Goldsmith, C.D., and Hater, G.R. 2004. Evaluation of a biologically active cover for mitigation of landfill gas emissions. *Environmental Science and Technology*, 38, 4891–4899.

Barlaz, M.A., Ham, R.K., and Schaefer, D.M. 1990. Methane production from municipal refuse: A review of enhancement techniques and microbial dynamics. *Critical Review in Environmental Control*, 19(6), 557–584.

Bast, H., Funke, S., Sanders, P., and Schultes, D. 2007. Fast routing in road networks with transit nodes. *Science*, 316(5824), 566.

Bastida, F., Kandeler, E., Moren, J.L., Ros, M., Garcia, C., and Hernandez, T. 2008. Application of fresh and composted organic wastes modifies structure, size and activity of soil microbial community under semiarid climate. *Applied Soil Ecology*, 40, 318–329.

Belyaeva, O.N. and Haynes, R.J. 2010. A comparison of the properties of manufactures soils produced from composting municipal green waste alone or with poultry manure or grease trap/septage waste. *Biology and Fertility of Soil*, 46, 271–281.

Bhambulkar, A.V. and Khedikar, I.P. 2011. Municipal solid waste (MSW) collection route for Laxmi Nagar by geographical information system (GIS). *International Journal of Advanced Engineering Technology*, 2(4), 48–53.

Bhawalkar, U.S. and Bhawalkar, U.V. 1993. Vermi-culture biotechnology. In P.K. Thampan (Ed.), *Organic in Soil Health and Crop Production*, pp. 69–85. Cochin, India: Peekay Tree Crops Development Foundation.

Bhide, A.D. and Shekdar, A.V. 1998. Solid waste management in Indian urban centers. *International Solid Waste Association Times (ISWA)*, 1, 26–28.

Bizukojc, E.L., Bizukojc, M., and Ledakowicz, S. 2002. Kinetics of the aerobic biological degradation of shredded municipal solid waste in liquid phase. *Water Research*, 36, 2121–2132.

Boss, J.C., Poyyamoli, G., and Roy, G. 2013. Evaluation of biomedical waste management in the primary and community health centers in Puducherry region, India. *International Journal of Current Microbiology and Applied Sciences*, 2(12), 592–604.

Bundela, P.S., Gautam, S.P., Pandey, A.K., Awasthi, M.K. and Sarsaiya, S. 2010. Municipal solid waste management in Indian cities: A review. *International Journal of Environmental Sciences*, 1(4), 591–606.

Center for Environment and Development (CED). 2012. Short term course on solid waste management. http://cedindia.org/organization/affiliation/centre-of-excellence-of-moud-govt-of-india (accessed September 11, 2015).

Central Public Health and Environmental Engineering Organisation (CPHEEO). 2000. *Manual on Water Supply and Treatment*. http://cpheeo.nic.in/Watersupply.htm.

Chakrabarty, P., Srivastava, V.K., and Chakrabarti, S.N. 1995. Solid waste disposal and the environment – A review. *Indian Journal of Environmental Protection*, 15(1), 39–43.

Chang, J., Tsai, J.J., and Wu, K.H. 2006. Thermophilic composting of food waste. *Bioresource Technology*, 97, 116–122.

Chang, N.B. and Pires, A. 2015. Systems engineering principles for solid waste management. In N.B. Chang and A. Pires, *Sustainable Solid Waste Management: A System Engineering Approach*, First Edition, pp. 215–233. Hoboken, NJ: Wiley.

Chang, N.B., Pires, A., and Martinho, G. 2011. Empowering systems analysis for solid waste management: Challenges, trends and perspectives. *Critical Reviews in Environmental Science and Technology*, 41(16), 1449–1530.

Checkland, P. 2000. Soft systems methodology: A thirty year retrospective. *Systems Research and Behavioral Science*, 17(S1), 11–58.

Chefetz, B., Hatcher, P.G., Hadar, Y., and Chen, Y. 1998. Characterization of dissolved organic matter extracted from composted municipal solid waste. *Soil Science Society of American Journal*, 62, 326–332.

Chen, Y. and Inbar, U. 1993. Chemical and spectroscopical analyses of organic matter transformation during composting in relation to compost maturity. In H.A.J. Hoitink and H.M. Keener (Eds), *Science and Engineering of Composting: Design, Environmental, Microbiological and Utilization Aspects*, pp. 550–600. Worthington, OH: Renaissance Publications.

Christensen, T.H., Cossu, R., and Stegmann, R. 1989. *Sanitary Landfilling: Process, Technology and Environmental Impacts*, pp. 29–49. San Diego, CA: Academic Press.

Christensen, T.H. and Kjeldsen, P. 1989. Basic biochemical processes in landfills. In T.H. Christensen, R. Cossu, and R. Stegmann (Eds), *Sanitary Landfilling: Process, Technology, and Environmental Impact*, pp. 29–49. San Diego, CA: Academic Press.

Coffey, M. and Coad, A. 2010. *Collection of Municipal Solid Waste in Developing Countries*. Malta: UN-Habitat.

Cohen, B. 2004. Urban growth in developing countries: A review of current trends and a caution regarding existing forecasts. *World Development*, 32(1), 23–51.

Cointreau, S. 2006. Occupational and environmental health issues of solid waste management: Special emphasis on developing countries. Washington, DC: World Health Organization (WHO).

Colon, M. and Fawcett, B., 2006. Community-based household waste management: Lessons learnt from EXNOR's zero waste management scheme in two south Indian cities. *Journal of Habitat International*, 30(4), 916–931.

Consonni, S., Giugliano, M., Massarutto, A., Ragazzi, M., and Saccani, C. 2011. Material and energy recovery in integrated waste management systems: Project overview and main results. *Waste Management*, 31(9–10), 2057–2065.

Daskalopoulos, E. 1998. An integrated approach to municipal solid waste management. *Resources, Conservation and Recycling*, 24(1), 33–50.

Damghani, A.M., Savarypour, G., Zand, E., and Deihimfard, R. 2008. Municipal solid waste management in Tehran: Current practices, opportunities and challenges. *Waste Management*, 28, 929–934.

Dangi, M.B., Pretz, C.R., Urynowicz, M.A., Gerow, K.G., and Reddy, J.M. 2011. Corrigendum to municipal solid waste generation in Kathmandu, Nepal. *Journal of Environment Management*, 91, 240–249.

Das, S. 2014. Estimation of municipal solid waste generation and future trends in greater metropolitan regions of Kolkata India. *Journal of Industrial Engineering and Management Innovation*, 1, 31–38.

Das, S., Bhattacharyya, B.K. 2015. Optimization of municipal solid waste collection and transportation routes. *Waste Management*, 43, 9–18.

Davis, M.E. and Davis, R.J. 2003. Reactors for measuring reactor rates. In *Fundamentals of Chemical Reaction Engineering*, pp. 64–100. Boston, MA: McGraw-Hill.

Delgado-Moreno, L., Pena, A., and Mingorance, M.D. 2009. Design of experiments in environmental chemistry studies: Example of the extraction of trianzine from soil after olive cake amendment. *Journal of Hazardous Material*, 162, 1121–1128.

Derya, A., Agdag, N.O., and Simsek, O. 2013. Seed sludge effect on anaerobic treatment of organic solid waste and its microbial community. In *Digital Proceeding of the ICOEST*. pp. 300–310. Nevsihir, Turkey.

Dev, S.M. and Yedla, S. 2015. *Cities and Sustainability Issues and Strategic Planning*. New Delhi, India: Springer.

Dijkstra, E.W. 1959. A note on two problems in connexion with graphs. *Numerische Mathematik*, 1, 269–271.

Donovan, S.M., Bateson, T., Gronow, J.R., and Voulvoulis, N. 2010. Modelling the behaviour of mechanical biological treatment outputs in landfills using the GasSim model. *Science of the Total Environment*, 408(8), 1979–1984.

Dyson, B. and Chang, N.B. 2005. Forecasting municipal solid waste generation in a fast-growing urban region with system dynamics modeling. *Waste Manage*, 25(7), 669–679.

El-Fadel, M., Findikakis, A.N., and Leckie, J.O. 1988. A numerical model for methane production in managed sanitary landfills. *Waste Management and Research*, 7, 31–42.

El-Fadel, M., Findikakis, A.N., and Leckie, J.O. 1996. Numerical modelling of generation and transport of gas and heat in landfills: Model formulation. *Waste Management and Research*, 14, 483–504.

Emcon Associates. 1980. *Methane Generation and Recovery from Landfills*, pp. 37–58. Ann Arbor, MI: Science Publishers.

Emery, A., Davies, A., Griffits, A., and William, K. 2007. Environmental and economic modeling: A case study of municipal solid waste management scenarios in waste management scenarios in whales. *Resources, Conservation and Recycling*, 49, 244–263.

ENVIS Centre on Municipal Solid Waste Management. 2016. Sanitary landfill. http://nswaienvis.nic.in/Waste_Portal/SanitaryLandfill.aspx (accessed January 20, 2016).

Eriksen, G., Coale, F., and Bollero, G. 1999. Soil nitrogen dynamics and maize production in municipal solid waste amended soil. *Agronomy Journal*, 91, 1009–1016.

Eriksson, O., Bisaillon, M., Haraldsson, M., and Sundberg, J. 2014. Integrated waste management as a mean to promote renewable energy. *Renewable Energy*, 61, 38–42.

Finstein, M.S. and Miller, F.C. 1985. Principal of composting leading to maximization of decomposition rate, odour control and cost effectiveness. In J.K.R. Gasser (Ed.). *Composting of Agricultural and Other Wastes: Seminar by the CEC, Brasenose College*, Oxford, 19–20 March 1984. London: Elsevier.

Fourti, O., Haydri, Y., Murano, F., Jedidi, N., and Hassen, A. 2008. A new process assessment of co-composting of municipal solid waste and sewage sludge in semi-arid pedo-climatic condition. *Annals of Microbiology*, 58(3), 403–409.

Fulekar, M.H., Pathak, B., and Kale, R.K. 2014. *Environment and Sustainable Development*. New Delhi, India: Springer.

Garcia-Gil, J.C., Plaza, C., Soler-Rovira, P., and Polo, A. 2000. Long-term effects of municipal solid waste compost application on soil enzyme activities and microbial biomass. *Soil Biology and Biochemistry*, 32, 1907–1913.

Garcia-Heras, J.L. 2003. Reactor sizing, process kinetics and modelling of anaerobic digestion of complex wastes. In J. Mata-Alvarez (Ed.), *Biomethanization of the Organic Fraction of Municipal Solid Waste*, pp. 21–62. Padstow, Cornwall: IWA Publishing, TJ International Ltd.

Geisberger, R. 2011. Advanced route planning in transportation networks. PhD thesis, Karlsruhe Institute of Technology, Karlsruhe, Germany.

Genyanst, J.S. and Lei, F. 2003. Microbial community structure dynamics during aerated and mixed composting. *American Society of Agricultural and Biological Engineers*, 46, 577–584.

Getahun, T.E., et al. 2012. Municipal solid waste generation in growing urban areas in Africa: Current practices and relation to socioeconomic factors in Jimma, Ethiopia. *Environmental Monitoring and Assessment*, 184(10), 63 37–45.

Ghaly, A.E., Alkoaik, F., and Snow, A. 2006. Thermal balance of in-vessel composting of tomato plant residues. *Canadian Biosystems Engineering*, 48, 1–11.

Ghosh, C. 2004. Integrated vermin-pisciculture: An alternative option for recycling of municipal solid waste in rural India. *Bioresource Technology*, 93(1), 71–75.

Global Waste Management Market Assessment Report. 2007. Key Note Publications. March 1, 2007. http://www.seas.columbia.edu/earth/wtert/sofos/Key_Global_Waste_Generation.pdf (accessed August 27, 2015).

Goldberg, A.V. and Harrelson, C. 2005. Computing the shortest path: A* search meets graph theory. In *Proceedings of the 16th Annual AC–SIAM Symposium on Discrete Algorithms (SODA '05)*, pp. 156–165. Philadelphia, PA: SIAM.

Goyal, D., Kumar, S., and Sil, A. 2014. Municipal solid waste: Zero tolerance management strategy. *International Journal of Environment Technology and Management*, 17(2–4), 113–121.

Graves, R.E., Hattemer, M.G., Stettler, D., Krider, J.N., and Chapman, D. 2010. Composting. In *Environmental Engineering National Engineering Handbook*, Part 637. 210-VI-NEH, Amend. 40, November 2010. Washington, DC: United States Department of Agriculture.

Gupta, T.N. 1998. *Building Materials in India: 50 Years; A Commemorative Volume.* New Delhi, India: Building Materials Technology Promotion Council, Government of India.

Gupta, S., Krishna, M., Prasad, R.K., Gupta, S., and Kansal, A. 1998. Solid waste management in India: options and opportunities. *Resource, Conservation and Recycling*, 24, 137–154.

Haan, H.C., Coad, A., and Lardinois, I. 1998. *Municipal Waste Management: Involving Micro- and Small Enterprises; Guidelines for Municipal Managers.* Turin, Italy: International Training Centre of the ILO, SKAT, WASTE.

Haarstrick, A., Hempel, D.C., Ostermann, L., Ahrens, H., and Dinkler, D. 2001. Modelling of the biodegradation of organic matter in municipal landfills. *Waste Management and Research*, 19(4), 320–331.

Hamoda, M.F., Abu Qdais, H.A., and Newham, J. 1998. Evaluation of municipal solid waste composting kinetics. *Resources Conservation and Recycling*, 23, 209–223.

Hardoy, J.E., Mitlin, D., Satterthwaite, D., and Hardoy, J.E. 2001. *Environmental Problems in an Urbanizing World: Finding Solutions for Cities in Africa, Asia, and Latin America.* Sterling, VA: Earthscan Publications.

Haug, R.T. 1993. *The Practical Handbook of Compost Engineering.* Boca Raton, FL: Lewis Publications.

He, Y., Inamori, Y., Mizuochi, M., Kong, H., Iwami, N., and Sun, T. 2001. Nitrous oxide emissions from aerated composting of organic waste. *Environment Science and Technology*, 35, 2347–2351.

Hecklinger, R.S. 1996. *The Engineering Handbook.* Boca Raton, FL: CRC Press/Taylor & Francis.

Heldivakis, C.P. 1983. Methanogenesis in solid waste landfill bioreactors. PhD thesis, Stanford University, Stanford, CA.

Hettiaratchi, J.P.A., Jayasinghe, P.A., Tay, J.H., and Yadev, S.K. 2015. Recent advances in biomass to energy using landfill bioreactor technology. *Current Organic Chemistry*, 19(5), 413–422.

Hilary, T., Tchobanoglous, G., and Kreith, F. 2002. *Handbook of Solid Waste Management*, Second Edition. Collection of solid waste. pp. 7.25–7.27. New York: McGraw-Hill.

Hoornweg, D. and Giannelli, N. 2007. Managing municipal solid waste in Latin America and the Caribbean: Integrating the private sector, harnessing incentives. https://openknowledge.worldbank.org/handle/10986/10639.

Hsu, J. and Lo, S. 1999. Chemical and spectroscopic analysis of organic matter transformations during composting of pig manure. *Environment Pollution*, 104, 189–196.

Huang, R.T. 1993. *Practical Handbook of Compost Engineering.* Boca Raton, FL: Lewis Publishers.

Huang, D., Zeng, G., Feng, C., Hu, S., Lai, C., Zhao, M., Su, F., Tang, L., and Liu, H. 2010. Change of microbial population structure related to lignin degradation during lignocellulose waste composting. *Bioresource Technology*, 101, 4062–4067.

Hunag, D., Zeng, G., Feng, C., Hu, S., Zhao, M., Lai, C., Zhang, Y., Jiang, Z., and Liu, H. 2010. Mycelial growth and solid-state fermentation of lignocellulose waste by white-rot fungus *Phanerochaete chrysosporium* under lead stress. *Chemosphere*, 81, 1091–1097.

Intergovernmental Panel on Climate Change (IPCC). 1997. *Revised 1996 IPCC Guidelines for National Greenhouse Inventories*. Paris, France: IPCC, IPCC/OECD/IEA.

Intergovernmental Panel on Climate Change (IPCC). 1998. *Good Practice Guidance and Uncertainty Management in National Greenhouse Gas Inventories*, Chapter 5.

Intergovernmental Panel on Climate Change (IPCC). 2006. *IPCC Guidelines for National Greenhouse Gas Inventories*. Hayama, Japan: IPCC, IPCC/OECD/IEA/IGES.

Ionescu, R.D., Ragazzi, M., Battisti, L., Rada, E.C., and Ionescu, G. 2013. Potential of electricity generation from renewable energy sources in standard domestic houses. *WIT Transactions on Ecology and the Environment*, 176, 245–253.

Jafari, A., Godini, H., and Mirhousaini, H.S. 2010. Municipal solid waste management in Khoramabad city and experiences. *International Journal of Environmental, Chemical, Ecological, Geological and Geophysical Engineering*, 4(2).

Jakobson, S.T. 1994. Aerobic decomposition of organic wastes. *Resources, Conservation and Recycling*, 12, 165–175.

Jayaram, V., Nigam, A., Welch, W.A., Miller, J.W., and Cocker III, D.R. 2011. Effectiveness of emission control technologies for auxiliary engines on ocean-going vessels. *Journal of the Air & Waste Management Association*, 61, 14–21.

Jayasinghe, P.A. 2013. Enhancing gas production in landfill bioreactors by leachate augmentation. PhD thesis, Department of Chemical and Petroleum Engineering, University of Calgary, Canada.

Jones, K.L. and Grainger, K.M. 1983. The application of enzyme activity measurements to a study of factors affecting protein, starch and cellulose fermentation in a domestic landfill. *European Journal of Applied Microbiology and Biotechnology*, 18, 181–189.

Kadafa, A.A. 2012. A review on municipal solid waste management in Nigeria. *Journal of American Science*, 8(12), 975–982.

Kalia, A.K. and Singh, S.P. 2001. Effect of mixing digested slurry on the rate of biogas production from dairy manure in batch fermenter. *Energy Sources*, 23, 711–715.

Kamalan, H., Sabour, M., and Shariatmadari, N. 2011. A review of available landfill gas models. *Journal of Environment Science and Technology*, 4(2), 79–92.

Kanamori, Y. 2012. Household consumption change and household waste generation from household activities in Asian countries. *Berlin Conference on the Human Dimensions of Global Environmental Change*.

Kansal, A., Prasad, R.K., and Gupta, S. 1998. Delhi municipal solid waste and environment – An appraisal. *Indian Journal of Environmental Protection*, 18(2), 123–128.

Karunarathne, L. 2015. Municipal solid waste management (MSWM) in Sri Lanka. *Proceedings of the National Symposium on Real Estate Management and Valuation*.

Katheem, K.S., Ibrahim, H.I., Quaik, S., and Ismail, A.S. 2016. *Prospects of Organic Waste Management and the Significance of Earthworms*. Cham, Switzerland: Springer International Publishing.

Khan, R.R. 1994. Environmental management of municipal solid wastes. *Indian Journal of Environmental Protection*, 14(1), 26–30.

Khanal, S.K. 2008. Microbiology and biochemistry of anaerobic biotechnology. In *Anaerobic Biotechnology for Bioenergy Production*, pp. 29–41. Ames, IA: Wiley-Blackwell.

Khatib, A.I. 2011. Municipal solid waste management in developing countries: Future challenges and possible opportunities. In S. Kumar (Ed.), *Integrated Waste Management*, 35–48.

Khwairakpam, M. and Bhargava, R. 2009. Bioconversion of filter mud using vermi-composting employing two exotic and one local earthworm species. *Bioresource Technology*, 100, 5846–5852.

Kim, D.H. and Oh, S.E. 2011. Continuous high-solids anaerobic co-digestion of organic solid wastes under mesophilic conditions. *Waste Management*, 31, 1943–1948.

Konteh, F.H. 2009. Urban sanitation and health in the developing world: Reminiscing the nineteenth century industrial nations. *Health and Place*, 15(1), 69–78.

Kumar, M. and Kumar, B.S. 2015. Municipal solid waste management in Coimbatore city. *Asian Journal of Biochemical and Pharmaceutical Research*, 1(5), 74–77.

Kumar, S., Gaikward, S.A., Shekdar, A.V., Kshirsagar, P.S., and Singh, R.N. 2004. Estimation method for national methane emission from solid waste landfills. *Atmospheric Environment*, 38, 3481–3487.

Lekasi, J.K., Tanner, J.C., Kimani, S.K., and Harris, P.J.C. 2003. Cattle manure quality in Maragua District, Central Kenya: Effect of management practices and development of simple methods of assessment. *Agricultural Ecosystems and Environment*, 94, 289–298.

Leskovac, V. 2003. *Comprehensive Enzyme Kinetics*. New York: Kluwer Academic Publishers.

Li, X., Xing, M., Yang, J., and Huang, Z. 2011. Compositional and functional features of humic acid-like fractions from vermin-composting of sewage sludge and cow dung. *Journal of Hazardous Material*, 185, 740–748.

Lin, Y.P., Huang, G.H., Lu, H.W., and He, L. 2008. Modeling of substrate degradation and oxygen consumption in waste composting processes. *Waste Management*, 28, 1375–1385.

Liptak, G.B. and Liu, H.F.D. 1999. *Solid Waste in Environmental Engineers' Handbook*. Boca Raton, FL: CRC Press.

Mahar, A., Malik, R.N., Qadir, A., Ahmed, T., Khan, Z., and Khan, M.A. 2007. Review and analysis of current solid waste management situation in urban areas of Pakistan. *Proceedings of the International Conference on Sustainable Solid Waste Management*, pp. 34–41. Chennai, India: Centre for Environmental Studies, Department of Civil Engineering, Anna University.

Maier, M.W. 1998. Architecting principles for system of systems. *Systems Engineering*, 1(4), 267–284.

Malviya, R., Chaudhary, R., and Buddhi, D. 2002. Study on solid waste assessment and management – Indore city. *Indian Journal of Environmental Protection*, 22(8), 841–846.

Mane, A.V. and Praveen A. 2013. Municipal solid waste management: A case study of Phursungi plant, Pune. *World Journal of Environmental Biosciences*, 2, 89–99.

Marshall, E.R. and Bakhsh, K.F. 2013. Systems approaches to integrated solid waste management in developing countries. *Waste Management*, 33, 988–1003.

Marshall, E.R. and Bakhsh, K.F. 2015. System approach to integrated solid waste management in developing countries. *Waste Management*, 33, 988–1003.

Massarutto, A., Carli, A.D., and Graffi, M. 2011. Material and energy recovery in integrated waste management systems: A life-cycle costing approach. *Waste Management*, 31, 9–10.

Mata-Alvarez, J. 2003a. Anaerobic digestion of the organic fraction of municipal solid waste: A perspective. In J. Mata-Alvarez (Ed.), *Biomethanization of the Organic Fraction of Municipal Solid Waste*, pp. 90–105. London: IWA Publishing.

Mata-Alvarez, J. 2003b. Fundamentals of the anaerobic digestion process. In J. Mata-Alvarez (Ed.), *Biomethanization of the Organic Fraction of Municipal Solid Waste*, pp. 1–20. London: IWA Publishing.

Maudgal, S.C. 2011. Waste management in India. *Journal of Indian Association for Environmental Management*, 22, 203–208.

McCarty, P.L. 1964. Anaerobic waste treatment fundamentals. *Public Works*, 95, 91–126.

McDougall, F., White, P.R., Franke, M., and Hindle, P. 2001. *Integrated Solid Waste Management: A Lifecycle Inventory*, Second Edition. Oxford, UK: Blackwell Science.

McInerney, M.J., Bryant, M.P., and Pfennig, N. 1979. Anaerobic bacterium that degrades fatty acids in syntrophic association with methanogens. *Archives of Microbiology*, 122(2), 129–135.

Meadows, D.H. 2008. *Thinking in Systems: A Primer*. Sterling, VA: Earthscan.

Medina, M. 2000. Scavenger cooperatives in Asia and Latin America. *Resources, Conservation and Recycling*, 31(1), 51–69.

Medina, M. 2010. Solid wastes, poverty and the environment in developing country cities. Working paper, United Nations World Institute for Development Economic Research (accessed February 1, 2016).

Medina, M. and Dows, M. 2000. A short history of scavenging. *Comparative Civilizations Review*, 42, 7–17.

Memon, M.A. 2010. Integrated solid waste management based on the 3R approach. *Journal of Material Cycles and Waste Management*, 12(1), 30–40.

Ministry of Environment and Forests, Government of India. 2000. Notification. New Delhi, India. http://www.moef.nic.in/legis/hsm/mswmhr.html (accessed May 27, 2016).

Ministry of Environment, Forest and Climate Change, Government of India. 2015. Draft rules. New Delhi, India. http://www.moef.nic.in/sites/default/files/SWM%20Rules%202015%20-Vetted%201%20-%20final.pdf (accessed May 27, 2016).

Missen, R.W., Mims, C.A., and Saville, B.A. 1999. Biochemical reactions: Enzyme kinetics. In *Introduction to Chemical Reaction Engineering and Kinetics*, pp. 261–278. New York: Wiley.

Monnet, F. 2003. An introduction to anaerobic digestion of organic wastes. Remade Scotland. http://www.remade.org.uk/media/9102/an%20introduction%20to%20anaerobic%20digestion%20nov%202003.pdf (accessed October 15, 2015).

Mor, S., Ravindra, K., Visscher, A.D., Dahiya, R.P., and Chandra, A. 2006. Municipal solid waste characterization and its assessment for potential methane generation: a case study. *Journal of Science of the Total Environment*, 371(1), 1–10.

Mormile, M.R., Gurijala, K.R., Robinson, J.A., McInerney, M.J., and Suflita, J.M. 1996. The importance of hydrogen in landfill fermentations. *Applied Environmental Microbiology*, 62(5), 1583–1588.

Morrissey, A.J. and Browne, J. 2004. Waste management models and their application to sustainable waste management. *Waste Management*, 24(3), 297–308.

Mrówczyńska, B. 2014. Route planning of separate waste collection on a small settlement. *Transport Problems*, 9(1), 61–68.

Muzenda, E., Ntuli, F., and Pilusa, J.T. 2012. Waste management, strategies and situation in South Africa: An overview. *Waste Management*, 6(8), 552–555.

Nakasaki, K., Tran, L.T.H., Idemoto, Y., Abe, M., and Rollon, A.P. 2009. Comparison of organic matter degradation and microbial community during thermophilic composting of two different types of anaerobic sludge. *Bioresource Technology*, 100, 676–682.

Nema, A.K. 2004. *Collection and Transport of Municipal Solid Waste*. In Training program on solid waste management. Delhi, India: Springer.

Ojha, A., Reuben, C.A., and Sharma, D. 2012. Solid waste management in developing countries through plasma arc gasification: An alternative approach. *APCBEE Procedia*, 1, 193–198.

Olaosebikan, K.O., Salami, L., Adeoye, K.B., and Ajayi, O.T. 2012. Achieving sustainable development in small communities via combined heat and power. *Scientific Research*, 4, 160–169.

Olley, J., Scheinberg, A., Wilson, D., and Read, A. 2003. Building stakeholder capacity for integrated sustainable waste management planning. Skat Foundation workshop on "Solid waste collection that benefits the urban poor," Dar es Salaam, Tanzania.

Pakistan Environmental Protection Agency (Pak EPA). 2005. Guidelines for solid waste management (draft). http://environment.gov.pk/EA-GLines/SWMGLinesDraft.pdf (accessed November 10, 2015).

Pappu, A., Saxena, M., and Asolekar, S.R. 2007. Solid wastes generations in India and their recycling potential in building materials. *Building and Environment*, 42, 2311–2320.

Pachauri, R.K. and Batra R.K. 2001. *Directions, Innovations and Strategies for Harnessing Action for Sustainable Development*, New Delhi, India: TERI.

Pawale, D.P., Kadam, S.M., Sarawade, S.S., and Pawar, S.M. 2015. Design of hybrid dryer for municipal solid waste. *Golden Research Thoughts*, 4(10), 1–31.

Pires, A., Martinho, G., and Chang, N.B. 2011. Solid waste management in European countries: A review of systems analysis techniques. *Journal of Environmental Management*, 92, 1033–1050.

Purohit, H.J., Raje, D.V., Kapley, A., Padmanabhan, P., and Singh, R.N. 2003. Genomics tools in environment impact assessment. *Environmental Science and Technology*, 37, 356–363.

Qdais, H.A. and Alsheraideh, A.A. 2008. Kinetics of solid waste biodegradation in laboratory lysimeters. *Jordan Journal of Civil Engineering*, 2(1), 45–52.

Ramachandra, T.V. 2006. Management of municipal solid waste, Commonwealth of Learning Indian Institute of Science TERI Press. https://books.google.co.in/books?hl=en&lr=&id=H1l-QWqSJwcC&oi=fnd&pg=PR11&dq=ramachandra+2000+solid+waste+management&ots=Z3ud3QGnwE&sig=Gl5x58dFuvUifgEC_qgYOzTLiTY#v=onepage&q=ramachandra%202000%20solid%20waste%20management&f=false (accessed on December 18, 2015).

Ramachandra, T.V. 2011. Integrated management of municipal solid waste. In S.R. Garg (Ed.), *Environmental Security: Human and Animal Health*, pp. 465–484. Lucknow, India: IBDC.

Ramasamy, K. and Nagamani, B. 2010. Biogas production technology: An Indian perspective. Department of Environmental Sciences, Tamil Nadu Agricultural University. http://www.iisc.ernet.in/currsci/jul10/articles13.htm (accessed October 22, 2015).

Reeb, J. and Milota, M. 1999. Moisture content by the oven dry method for industrial testing. *WDKA*, May 1999, 1(1–3).

Rotten Truth: About Garbage. n.d. A garbage timeline. Association of Science-Technology Centers Incorporated and the Smithsonian Institution Traveling Exhibition Service. www.astc.org/exhibitions/rotten/timeline.htm.

Rynk, R. et al. 1992. In R. Rynk (Ed.), Benefits and drawbacks, characteristics of raw materials. *On-Farm Composting Handbook*, pp. 6–13, 106–113. Ithaca, NY: Northeast Regional Agricultural Engineering Service.

Salvato, A.J. 1992. *Environmental Engineering and Sanitation*, pp. 729–764. New York: Wiley-Interscience Publication, John Wiley.

Sanders, P. and Schultes, D. 2006. Engineering highway hierarchies. In *Proceedings of the 14th Annual European Symposium on Algorithms (ESA'06)*, pp. 804–816. Lecture Notes in Computer Science, Vol. 4168. Berlin, Germany: Springer.

Sanders, W.T.M., Cveeken, A.H.M., Zeeman, G., and VanLier, J.B. 2003. Analysis and optimization of the anaerobic digestion of the organic fraction of municipal solid waste. In J. Mata-Alvarez (Ed.), *Biomethanization of the Organic Fraction of Municipal Solid Waste*, pp. 63–89. London: IWA Publishing.

Scheinberg, A. 2001. Integrated Sustainable Solid Waste Management. The Concepts, Tools for Decision-makers. Experiences from Urban Waste Expertise Programme (1995–2001). http://www.waste.nl/sites/waste.nl/files/product/files/tools_finecon_eng1.pdf (accessed on Decemebr 5, 2015).

Scherer, P.A., Vollmer, G.R., Fakhouri, T., and Martensen, S. 2001. Development of a methanogenic process to degrade exhaustively the organic fraction of municipal "grey waste" under thermophilic and hyperthermophilic conditions. *Water Science and Technology*, 41(3), 83–91.

Scheutz, C., et al. 2011. Gas production composition and emission at a modern disposal site receiving waste with a low-organic content. *Waste Management*, 31(5) 946–955.

Schubeler, P., Wehrle, K., and Christen, J. 1996. *Conceptual Framework for Municipal Solid Management in Low Income Countries*. St Gallen, Switzerland: Swiss Centre for Development Corporation in Technology and Management.

Schwarz-Herion, O., Omran, A., and Rapp, H.P. 2008. A case study on successful municipal solid waste management in industrialized countries by the example of Karlsruhe city, Germany. *Journal of Engineering*, 6(3), 266–273.

Scottish Environmental Protection Agency. 2002. Guidance on landfill gas flaring. https://www.gov.uk/government/uploads/system/uploads/attachment_data/file/321623/Guidance_on_landfill_gas_flaring.pdf (accessed January 20, 2016).

Seadon, J.K. 2010. Sustainable waste management systems. *Journal of Cleaner Production*, 18(16–17), 1639–1651.

Sen, B. and Chandra, T.S. 2006. Chemolytic and solid-state spectroscopic evaluation of organic matter transformation during vermin-composting of sugar industry waste. *Bioresource Technology*, 98(8), 1680–1683.

Sengupta, J. 2002. Recycling of agro-industrial wastes for manufacturing of building materials and components in India: An overview. *Civil Engineering and Construction Review*, 2, 23–33.

Sharholy, M., Ahmad, K., Mahmood, G., and Trivedi, R.C. 2006. Development of prediction models for municipal solid waste generation for Delhi city. *Proceedings of National Conference of Advanced in Mechanical Engineering*, 1176–1186.

Sharholy, M., Ahmad, K., Vaishya, R.C., and Gupta, R.D. 2007. Municipal solid waste characteristics and management in Allahabad, India. *Journal of Waste Management*, 27(4), 490–496.

Shekdar, V.A. 2009. Sustainable solid waste management: An integrated approach for Asian countries. *Waste Management*, 29, 1438–1448.

Shermann-Huntoon, R. 2000. Latest developments in mid–large scale vermicomposting. *Biocycle Magazine*, November, p. 51. http://wormswork.pbworks.com/w/page/11865851/Latest%20Developments%20in%20Mid-Large%20Scale%20Vermicomposting (accessed December 31, 2015).

Shukor, A.S.F., Mohammed, H.A., Sani, H.A., and Awang, A. 2011. A review on the success factors for community participation in solid waste management. *Proceedings of the International Conference on Management*, pp. 963–976. June 13–14, Penang, Malaysia.

Siddiqui, T.Z., Siddiqui, F.Z., and Khan, E. 2006. Sustainable development through integrated municipal solid waste management (MSWM) approach – A case study of Aligarh District. *Proceedings of National Conference of Advanced in Mechanical Engineering*, 1168–1175.

Sil, A., Kumar, S., and Wong, J. 2014. Development of correction factors for landfill gas emission model suiting Indian condition to predict methane emission from landfills. *Bioresource Technology*, 168, 97–99.

Sivakumar, K. and Sugirtharan, M. 2010. Impact of family income and size on per capita solid waste generation: A case study in Manmunai North Divisional Secretariat division of Batticaloa. *Journal of Science University Kelaniya*, 5, 13–23.

Slezak, R., Krzystek, L., and Ledakowicz, S. 2012. Mathematical model of aerobic stabilization of old landfills. *Chemical Papers*, 66, 543–549.

Smith, J.M. 1981. *Chemical Engineering Kinetics*, Third Edition. McGraw-Hill Chemical Engineering Series. New York: McGraw-Hill.

Spellman, F.R. and Whiting, N.E. 2007. *Environmental Management of Concentrated Animal Feeding Operations (CAFOs)*. Boca Raton: CRC Press.

Spokas, K., Bogner, J., Chanton, J.P., Morcet, M., Aran, C., Graff, C., Moreau-Le Goluan, Y., and Hebe, I. 2006. Methane mass balance at three landfill sites: What is the efficiency of capture by gas collection systems? *Waste Management*, 26(5), 516–525.

Stanier, R.Y., Ingraham, J.L., Wheelis, M.L., and Painter, P.R. 1986. *The Microbial World*, Fifth Edition. Englewood Cliffs, NJ: Prentice-Hall.

Sternenfels, U.M.C. 2012. Compost physicochemical characteristics influencing methane biofiltration. PhD thesis, Department of Civil Engineering, University of Calgary, Canada.

Syeeda, A.U. and Rav, S.B. 2013. *Sustainable Solid Waste Management*. Oakville, ON: Apple Academic Press.

Tavares, G., Zsigraiova, Z., Semiao, V., and Carvalho, M.G. 2009. Optimization of MSW collection routes for minimum fuel consumption using 3D GIS modeling. *Waste Management*, 29, 1176–1185.

Tchobanoglous, G., Theisen, H., and Vigil, S. 1993. *Integrated Solid Waste Management Engineering Principles and Management Issues*. New York: McGraw-Hill.

Teodorita, A.S. et al. 2008. *Biogas Handbook*. Esbjerg, Denmark: BiG>East Project. http://www.lemvigbiogas.com (accessed October 20, 2015).

Thomas, B., Tamblyn, D., and Baetz, B. 1990. Expert systems in municipal solid waste management planning. *Journal of Urban Planning and Development*, 116, 150–155.

Thompson, S., Sawyer, J., Bonam, R., and Valdivia, J.E. 2009. Building a better methane generation model: Validating models with methane recovery rates from 35 Canadian landfills. *Waste Management*, 29, 2085–2091.

Timilsina, B.P. 2001. Public and private sector involvement in municipal solid waste management: An overview of strategy, policy and practices. *A Journal of the Environment,* 7, 68–77.

Tiquia, S.M. 2002. Evolution of extracellular enzyme activities during manure composting. *Journal of Applied Microbiology,* 92, 764–775.

Tiquia, S.M., Tam, N.F.Y., and Hodgkiss, I.S. 1996. Microbial activities during composting of spent pig manure saw dust litter at different moisture contents. *Bioresource Technology,* 55, 201–206.

Tongnetti, C., Mazzarino, M.J., and Laos, F. 2007. Co-composting biosolids and municipal organic waste: Effect of process management on stabilization and quality. *Biology and Fertility of Soils,* 43, 387–397.

Trautmann, N.M. and Krasny, M.E. 1997. *Composting in the Classroom,* The science of composting. pp. 1–26. Ithaca, NY: Cornell University.

Trihadiningrum, Y., Laksono, I.J., Dhokhikah, Y., Moesriati, A., Radita, R.D., and Sunaryo, S. 2015. Community activities in residential solid waste reduction in Tenggilis Mejoyo District, Surabaya City, Indonesia. *Journal of Material Cycles and Waste Management,* 1–10.

Troschinetz, M.A. and Mihelcic R.J. 2009. Sustainable recycling of municipal solid waste in developing countries. *Waste Management,* 29(2), 915–923.

Turner, R.K. and Powell, J.C. 1991. Towards an integrated waste management strategy. *Environmental Management and Health,* 2, 6–12.

UNDESA. 2005. Agenda 21 – Chapter 21 Environmentally Sound Management of Solid Wastes and Sewage-related Issues, Division for Sustainable Development, United Nations Department of Economic and Social Affairs (retrieved July 1, 2005).

UN-Habitat. 2010. *Solid Waste Management in the World's Cities: Water and Sanitation in the World's Cities.* London: Earthscan. http://mirror.unhabitat. org/pmss/listItemDetails.aspx?publicationID=2918 (accessed January 26, 2016).

United Nations Centre for Human Settlements. 1989. Community participation: Solid waste management in low-income housing projects; The scope for community participation. Metropolitan planning and management in developing world. Report, Nairobi.

United Nations Development Programme (UNDP). 2010. *Human Development Report.* New York: Palgrave Macmillan.

United Nations Environment Programme (UNEP). 2001. Annual Evaluation Report on *"Evaluation and Oversight Unit,"* pp. 32–38. http://www.unep.org/eou/ Portals/52/Reports/AnnualEvalReportEnglish2001.pdf (accessed August 8, 2015).

United Nations Environment Programme (UNEP). 2005. Sanitary Landfill. In *Solid Waste Management,* pp. 323–438. http://www.unep.org/ietc/Portals/136/SWM-Vol1-Part3.pdf (accessed November 21, 2015).

United Nations Environment Programme (UNEP). 2009. *Developing Integrated Solid Waste Management Plan Training Manual,* Vol. 2. Osaka, Japan. Assessment of Current Waste Management System and Gaps Therein. United Nations Environmental Programme Division of Technology, Industry and Economics International Environmental Technology Centre. http://www.unep.org/ ietc/Portals/136/Publications/Waste%20Management/ISWMPlan_Vol2.pdf (accessed September 12, 2015).

United Nations Environment Programme (UNEP). 2012. *Developing Integrated Solid Waste Management Plan: Training Manual*, Vol. 4, pp. 66–88. Osaka, Japan: UNEP, Division of Technology, Industry and Economics International Environmental Technology Centre.

United States Environmental Protection Agency (USEPA). 2005. Sanitary Landfill. In *Solid Waste Management*, pp. 323–438. http://www.unep.org/ietc/Portals/136/SWMVol1-Part3.pdf (accessed November 21, 2015).

United States Environmental Protection Agency (USEPA). 2011. *Inventory of US Greenhouse Gas Emissions and Sinks: 1990–2008*. Washington, DC: USEPA. http://www3.epa.gov/climatechange/Downloads/ghgemissions/US-GHG-Inventory-2011-Complete_Report.pdf (accessed December 11, 2015).

Unnikrishnan, S. and Singh, A. 2010. Energy recovery in solid waste management through CDM in India and other countries. *Resources, Conservation and Recycling*, 54(10), 630–640.

Upadhay, V., Jethoo, A.S., and Poonia, M.P. 2012. Solid waste collection and segregation: A case study of MNIT campus, Jaipur. *International Journal of Engineering and Innovative Technology (IJEIT)*, 1(3), 144–149.

Vavilin, V.A., Fernandez, B., Palatsi, J., and Flotats, X. 2008. Hydrolysis kinetics in anaerobic degradation of particulateorganic material: An overview. *Waste Management*, 28, 939.

Veenstra, S. 1997. Sanitation and solid waste technologies. Karunarathne, L. 2015. Municipal solid waste management (MSWM) in Sri Lanka. *Proceedings of the National Symposium on Real Estate Management and Valuation 2015*, Vol. 1.

Vesilind, P.A., Worrell, W.A., and Reinhart, D.R. 2002. *Solid Waste Engineering*. Pacific Grove, CA: Brooks/Cole.

Velzy, C.O. and Grillo, L.M., 2007. Waste-to-Energy combustion. In F. Kreith and D. Yogi (Ed.), *Handbook of Energy Efficiency and Renewable*, Boca Rotan: CRC Press.

Vidanaarachchi, C.K., Yuen, S.T.S., and Pilapitiya, S. 2006. Municipal solid waste management in the Southern Province of Sri Lanka: Problems, issues and challenges, *Waste Management*, 26, 920–930.

VNG International. 2010. *Municipal Development Strategy Process: A Toolkit for Practitioners*. The Hague, The Netherlands: VNG International.

Volokita, M., Adeliovich, A., and Soares, M.I.M. 2000. Detection of microorganism with overall cellulolytic activity. *Current Microbiology*, 40, 135–136.

Walling, E., Walston, A., Warren, E., Warshay, B., and Wilhelm, E. 2004. Municipal solid waste management in developing countries: Nigeria, a case study NTRES 314 Stephen Wolf 26 April 2004.

Wang, H. and Nie, Y. 2001. Municipal solid waste characteristics and management in China. *Journal of Air and Waste Management Association*, 51, 250–263.

Wang, L. 2014. *Sustainable Bioenergy Production*. Boca Raton, FL: CRC Press.

Wang, Q. 2004. Aspects of pretreated hospital waste biodegradation in landfills, pp. 71–85. PhD dissertation, Duisburg-Essen University, Germany.

Ward, A.J., Hobbs, P.J., Holliman, P.J., and Jones, D.L. 2008. Optimisation of the anaerobic digestion of agricultural resources. *Bioresource Technology*, 99, 7928–7940.

Warman, P.R. and Termeer, W.C. 2005. Evaluation of sewage sludge, septic waste and sludge compost applications to corn and forage: Ca, Mg, S, Fe, Mn, Cu, Zn and B content of crops and soil. *Bioresource Technology*, 96, 1029–1038.

WastePortal.net. Municipal strategic planning for solid waste management: Concepts and tools. http://wasteportal.net/en/waste-aspects/environmental-and-health-aspects/municipal-strategic-planning-solid-waste-management-c (accessed May 26, 2016).

Weng, Y.C. and Fujiwara, T. 2011. Examining the effectiveness of municipal solid waste management systems: An integrated cost–benefit analysis perspective with a financial cost modelling in Taiwan. *Waste Management*, 31(6), 1393–1406.

Weppen, P. 2001. Process calorimetry on composting of municipal organic wastes. *Journal of Biomass and Bioenergy*, 21, 289–299.

Waste-To-Energy Research and Technology Council (WTERT). 2012. WTERT 2012 Bi-Annual Conference. Columbia University. http://www.seas.columbia.edu/earth/wtert/sofos/WTERT2012Program.pdf.

White, C., Sayer, J.A., and Gadd, G.M. 1997. Microbial solubilization and immobilization of toxic metals: Key biogeochemical processes for treatments of contaminations. *FEMS Microbiology*, 20, 503–516.

White, D., Mottershead, D.N., and Harrison, S.J. 1984. Environmental Systems: An Introductory Text, Chapman and Wilson, D.C., 2007. Development Drivers for Waste Management. *Waste Management & Research*, 25 (3), pp. 198–207.

Williams, T.P. 2005. *Waste Treatment and Disposal*, pp. 325–342. New York: Wiley.

Wilson, D., Velis, C., and Cheeseman, C. 2006. Role of informal sector recycling in waste management in developing countries. *Habitat International*, 30, 797–808.

Wilson, D., Whiteman, A., and Tormin, A. 2001. Strategic planning guide for municipal solid waste management. Washington, DC World Bank. http://www.worldbank.org/urban/solid_wm/erm/start_up.pdf (accessed May 27, 2016).

Wilson, D.C. 2007. Development drivers for waste management. *Waste Management and Research*, 25(3), 198–207.

Wong, J.W.C., Fung, S.O., and Selvam, A. 2009. Coal fly ash and lime addition enhances the rate and efficiency of decomposition of food waste during composting. *Bioresource Technology*, 100, 3324–3331.

Wong, J.W.C., Mak, K.F., Chan, N.W., Lam, A., Fang, M., Zhou, L.X., Wu, Q.T., and Liao, X.D. 2001. Co-composting of soybean residues and leaves in Hong Kong. *Bioresource Technology*, 76, 99–106.

Woodruff. A. 2014. *Solid Waste Management in the Pacific Appropriate Technologies Asian Development Bank*. http://hdl.handle.net/11540/411.

World Bank. 2010. Waste generation. Urban Development Series: Knowledge Papers. http://siteresources.worldbank.org/INTURBANDEVELOPMENT/Resources/336387-1334852610766/Chap3.pdf (accessed May 27, 2016).

World Bank. 2012. *What a Waste: A Global Review of Solid Waste Management*. Washington, DC: World Bank.

Xing, M., Li, X., Yang, J., Huang, Z., and Lu, Y. 2012. Changes in the chemical characteristics of water-extracted organic matter from vermin-composting of sewage sludge and cow dung. *Journal of Hazardous Material*, 205–206, 24–31.

Yadav, A. and Garg, V.K. 2009. Feasibility of nutrient recovery from industrial sludge by vermicomposting technology. *Journal of Hazardous Material*, 168, 262–268.

Yadav, A. and Garg, V.K. 2011. Vermicomposting-An effective tool for the management of invasive weed *Parthenium hysterophorus*. *Bioresource Technology*, 102, 5891–5895.

Yoshida, M. and Sakurai, K. 2000. Municipal solid waste management in developing countries: towards more efficient international cooperation. *Waste Management Research*, 11(2), 142–151.

Yukon Environment. 2013. Construction requirements for new public waste disposal facilities. http://www.env.gov.yk.ca/air-water-waste/documents/solw6_construction_requirements_2013.pdf (accessed January 22, 2016).

Zahur, M. 2007. Solid waste management of Dhaka city. *Public Private Community Partnership*, 2, 93–97.

Zehnder, A.J.B. 1978. Ecology of methane formation. In R. Mitchell (Ed.), *Water Pollution Microbiology*, Vol. 2, pp. 349–376. New York: Wiley.

Zurbrugg, C., Drescher, S., Patel, A., and Sharatchandra, H.C. 2004. Decentralized composting of urban waste: An overview of community and private initiatives in Indian cities. *Waste Management*, 24, 655–662.

Index

A

Acclimation period, 99
Acetogenesis, 73–74, 94–95
Acidogenesis, 73
AD, *see* Anaerobic digester (AD)
Aeration process, 65
Aerobic composting, 60
Aerobic respiration, 60
Aerobic waste degradation process, 89–93
    factors affecting, 92–93
        feedstock composition, 92
        moisture, 92
        oxygen requirements, 92–93
        pH, 93
    initial mesophilic phase, 91
    kinetic model, 109–110
    secondary mesophilic phase, 91
    stoichiometric equation for, 91–92
    thermophilic phase, 91
Agricultural waste, 32
Amendment, compost mix, 61
Anaerobic degradation, 71
Anaerobic digester (AD)
    kinetics, 115–117
        batch, 116
        continuous flow, 116
        semibatch, 117
    waste degradation in, 100–101
        feedstock/substrate delivery
            controlled, 101
        moisture controlled, 101
        temperature controlled, 101
Anaerobic digestion
    biochemical processes of, 72–74
    parameters affecting, 74
    process, 71–72
        dry, 71
        wet, 71–72
    types of, 75
Anaerobic respiration, 60
Anaerobic waste degradation process,
        89, 93–99
    acetogenesis stage, 94–95

    factors affecting, 97–99
        inhibitors, 98–99
        moisture content, 98
        nutrients, 98
        pH, 97–98
        temperature, 98
    hydrolysis stage, 93–94
    kinetic model, 102–109
        formation of methane, 108–109
        intermediate, 106–108
        solid hydrolysis, 103–105
    methanogenesis stage, 95–96
    stoichiometric equation for, and
        estimation of theoretical
        methane yield, 96–97
APO, *see* Asian Productivity
        Organization (APO)
Aqueous carbon, 106–108
Area/ramp method, 85
Asian Development Bank Report, 2
Asian Productivity Organization
        (APO), 38
Attitude, people, 34–35

B

Batch digester, 116
Biochemical methane potential (BMP)
        assays, 150
Biodegradable waste, 145
Biofilters, 66
Biogas, 70, 73–74, 87–88
Biological processing, of municipal solid
        waste, 58–75
    biomethanation, 70–75
        biochemical processes of
            anaerobic digestion, 72–74
        parameters affecting anaerobic
            digestion, 74
        types of anaerobic digester, 75
    composting, 58–71
        chemical transformations
            during, 60

components of compost mix, 61
design of compost mixtures,
   60–61
main types of, system, 66–71
monitoring and parameter
   adjustment, 63–66
types of feedstock, 61–63
Biomass, 103, 106, 108
Biomedical waste, 146
Biomethanation, 71–75
   biochemical processes of anaerobic
      digestion, 73–74
   parameters affecting anaerobic
      digestion, 74–75
   types of anaerobic digester, 75
Biowaste bins, 147
BMP, *see* Biochemical methane potential
   (BMP) assays
Bottom-up design, 129
Brandling worms, *see Eisenia fetida*
Bulking agent, 61
Buswell equation, 96

C

C&D, *see* Construction and demolition
   (C&D) waste
Carbohydrates, 15
Carbon storage, in landfill, 88
Carbon-to-nitrogen (C:N) ratio, 92
Cause-and-effect concept, 126
CDM, *see* Clean development
   mechanism (CDM)
Centralized, and decentralized
   systems, 130
Central Pollution Control Board, 48
Chadwick, Edwin, 17
Chaetomium thermophilum, 63
Chemical characteristics, 15
Chemical oxygen demand (COD),
   98, 106
Citywide broadway concept, 18
Clean development mechanism
   (CDM), 150
Climate change, 122
Closed systems, 128
"Closing the loop", 126
Co-composting, 58

COD, *see* Chemical oxygen demand
   (COD)
Collection, of waste, 39
Commercial waste, 11, 29
Community activities, 34
Community bin collection systems,
   55–56, 145–147
Complex systems theory, 127
Compost, 58–59, 64–65, 68, 89, 92
   components of, mix, 60–61
   design of, mixtures, 60
Composting, 21, 58–71, 82
   chemical transformations
      during, 60
   components of compost mix, 60–61
   design of compost mixtures, 60
   main types of, system, 66–70
      in-vessel, 68
      vermicomposting, 68–71
      windrows, 67
   monitoring and parameter
      adjustment, 63–66
   types of feedstock, 61–62
Construction and demolition (C&D)
   waste, 31
Continuous flow digester, 116
Continuous stirred tank reactor (CSTR),
   116
Cultural, and socioeconomic aspects,
   123–124

D

Decomposition process, 88
Degradable organic carbon (DOC), 113
Density, of waste, 15
Destructive distillation, 80
Developed, and developing countries
   in integrated management of MSW
      sector, 8
   available technologies used in,
      23–25
Developing countries
   modern MSWM techniques in, 18–27
   available for implementation of, 20
   categories of problems, 26
   problems faced while
      implementing, 25–26

status of, with recent available
    technologies, 20–23
technologies used in, and
    developed countries, 23–25
MSW
    vs. developed countries in
        integrated management of, 8
    generation scenarios in, 36–38
    management practices of, 39–40
    on SWM, 5
Disciplinarity approach, 126
Disease Prevention Act, 18
Disposal, of solid wastes, 48–49
DOC, *see* Degradable organic
    carbon (DOC)
Domestic waste, *see* Household waste
Draft strategy, 140
Dry anaerobic process, 72
Dump site waste pickers, 51
Dust bin concept, 18
Dust yard system, 17

E

Earthworm castings, 69
Economic considerations, 84
Economic level, of sectors, 34
Eisenia fetida, 68
Environmental concerns, 121
Environmental factors, 84
Environmental management, 124
Environmental Protection Agency
    (EPA), 36
Enzyme-catalyzed kinetics, 104–105
EPA, *see* Environmental Protection
    Agency (EPA)
Extended producer responsibility
    (EPR), 22

F

Feedstock composition, 92
First-order waste hydrolysis kinetics,
    103–104
Fixed carbon, 16
Fluidized-bed incineration, 77, 78–79
FOD, *see* IPCC First-Order Decay (FOD)
    model

Forced aeration, 68
4R policies, and strategies, 6
Fukuoka University, 20

G

Garbage management, *see* Solid waste
    management
General systems theory (GST), 128
Geographic information system (GIS),
    136, 147
Ghana, 36, 38
GIS, *see* Geographic information
    system (GIS)
Global warming potential (GWP), 88
GOI, *see* Government of India (GOI)
Governance, concept of, 124
Government of India (GOI), 40, 134
GST, *see* General systems theory (GST)
GWP, *see* Global warming
    potential (GWP)

H

Half-life, 112
Hard systems thinking, 128
Heating value, 16
High-income countries, solid waste
    management development in,
    120–122
Household waste, 11, 29
House-to-house collection, 54–55
Human Development Report, 2
Hydrolysis, 73, 93–94

I

IFIs, *see* International financial
    institutions (IFIs)
IMSWM, *see* Integrated municipal solid
    waste management (IMSWM)
Incarnation, 82
Incineration, 21, 76–79
Industrialization, 33
Inert waste, 12, 145
Informal recycling systems, 51, 123
Inhibitors, 98–99
Initial mesophilic phase, 91

Inorganic waste, 14
Institutional waste, 29
Institutions, 124–125
Integrated municipal solid waste
    management (IMSWM), 23, 57;
    see also Municipal solid waste
    management (MSWM)
  factors affecting, 135
  generation-based, 137
  life cycle–based, 137
  management-based, 137
  planning, 135–136
Integrated solid waste management
    (ISWM), 5, 126
Interdisciplinary approach, 127
Intergovernmental Panel on Climate
    Change (IPCC), 113–114, 150
Intermediate anaerobic phase, 100
International financial institutions
    (IFIs), 125
International influences, 125
In-vessel composting, 68
IPCC, *see* Intergovernmental Panel on
    Climate Change (IPCC)
IPCC default method, 113–114, 150
IPCC First-Order Decay (FOD) model,
    113, 150
IPCC models, 148
Isolated system, 128
ISWM, *see* Integrated solid waste
    management (ISWM)

K

Kinetic model
  aerobic waste degradation, 109–110
  anaerobic waste degradation, 102–109
    formation of methane, 108–109
    intermediate, 106–108
    solid hydrolysis, 103–105

L

Landfill disposal site, *see* Material
    processing facility
Landfill gas (LFG) modeling, 148–151
Landfills
  carbon storage in, 88
  environmental impacts of, and their
    control, 83–84

gas generation kinetics, 110–115
  IPCC default method, 113–114
  IPCC First-Order Decay
    model, 113
  Scholl Canyon model and USEPA
    LandGEM, 110–112
  triangular model, 114–115
  sanitary, 84–88
  biogas recovery from, 87–88
  leachate collection system, 85–87
  methods of, 84–85
  types of, 82–83
    for milled waste, 83
    for mixed waste, 82–83
    monofills for specialized waste, 83
  waste degradation sequence in,
    99–100
LandGEM, *see* USEPA Landfill Gas
    Emissions Model (LandGEM)
LCA, *see* Life cycle assessment (LCA)
Leachate collection system, 85–87
Legislation and laws, in MSW, 40–49
LFG, *see* Landfill gas (LFG) modeling
Life cycle assessment (LCA), 134–135
Lipids, 15
Low- and medium-income countries,
    MSWM in, 122–125
  cultural and socioeconomic aspects,
    123–124
  international influences, 125
  political landscape, 124–125
  urbanization, 123
Lumbricus rubella, 69

M

Malaviya National Institute of
    Technology, 146
Mass balance approach, 110, 150
Mass-burn combustion, 77, 78
Material processing facility, 54
Material recovery facilities, 46
Mathematical model, 128
Matsufuji Yasushi, 20
Maturation phase, 100
MCs, *see* Municipal councils (MCs)
Meadows, D. H., 127
Memorandum of understanding
    (MoU), 139

Mesophilic microorganisms, 63
Methane (CH$_4$), 73, 108–109, 134, 150
Methanogenesis, 74, 95–96, 100
Michaelis–Menten model, 104, 105
Microbial growth, 107–108
Milled waste, landfills for, 83
Ministry of Environment and Forests
    (MoEF), 40
Ministry of Urban Development
    (MOUD), 134
Mixed waste, landfills for, 82–83
Mixing, 101
Moderate temperature phase, *see* initial
    mesophilic phase
Modern MSWM techniques, in
    developing countries, 18–27
  available for implementation of, 20
  categories of problems, 26
  problems faced while implementing,
    25–26
  status of, with recent available
    technologies, 20–23
  technologies used in developing and
    developed countries, 23–25
Modular incineration, 77, 78
MoEF, *see* Ministry of Environment and
    Forests (MoEF)
Moisture content, 14–15, 16, 61, 65, 92, 98
Monod equation, 107, 108
Monofills, for specialized waste, 83
Morris, Corbyn, 17
MoU, *see* Memorandum of
    understanding (MoU)
MOUD, *see* Ministry of Urban
    Development (MOUD)
MSW, *see* Municipal solid waste (MSW)
MSW Draft Rule 2015, 40–49
MSWM, *see* Municipal solid waste
    management (MSWM)
Multidisciplinarity approach, 126
Multistage digester, 75
Municipal corporation, role of, 49–50
Municipal councils (MCs), 40
Municipal solid waste (MSW), 1; *see also*
    Waste
  collection and transportation, 57
  composition and characteristics,
    12–16
    chemical, 15

physical, 14–15
proximate analysis, 16
ultimate analysis, 16
waste, 14
definition, 11
generation of, 29–38
  factors affecting rate of, 32–35
  scenarios in developing countries,
    36–38
  source of, 29–32
legislation and laws, in India, 40–49
  Draft Rule 2015, 40–49
management practices of, in
    developing countries, 39–40
mechanism of, generation, 11–12
need for integrated management of,
    4–8
  focus on, 6
  developing *vs.* developed
    countries in, 8
  research proving, 6–8
role of municipal corporation, 49–50
role of ragpickers in, 50–51
Municipal solid waste management
    (MSWM), 1
  component technologies for, 53–88
    biological and thermal processing
      methods, 58–81
    collection of, 54–55
    reuse and recycling, 80–81
    transportation, 56–58
    ultimate disposal methods, 81–88
  facets of, 3–4
  functional system, 17–27
    evolution of, 18
    historical development of, 17–18
    modern, techniques in developing
      countries, 18–27
  models for, 145–151
    community bin collection
      systems, 145–147
    landfill gas, 148–151
    vehicle routing, 147–148
  planning, 133–144
    effects of improper, for
      implementation of, 133–134
    long- and short-term, 140–144
    model, 144
    requirements in, 134–136

tactical and strategic, 137–140
systems approaches in, 119–132
  analysis techniques, 131–132
  centralized and decentralized
    systems, 130
  development drivers for, 120–125
  disciplinarity and
    multidisciplinarity, 126–127
  engineering principles, 127–129
  integrated, 126
  need for, 125–126
  system-of-systems, 129–130
urban problem, 2–3
"Municipal Solid Waste Management
  on a Regional Basis", 134
"Municipal Solid Waste Rules", 134
Municipal waste pickers, 51

N

National Policy of Solid Residues
  (NPSR), 36
NGOs, *see* Nongovernmental
  organizations (NGOs)
Nonbiodegradable waste, 145
Nongovernmental organizations
  (NGOs), 20, 49
Nonlandfill disposal approaches, 82
NPSR, *see* National Policy of Solid
  Residues (NPSR)
Nutrients, 74–75, 98

O

Occupational health risks, 51
Odor
  generation, 65–66
  management, 66
Open systems, 128
Organic loading rate, 75
Organics, waste, 12
Oxygen, 65, 92–93

P

pH, 64–65, 74, 93, 97–98
Physical characteristics, 14–15
Pluridisciplinary approach, 126–127
Policy measures, 124

Political landscape, 124–125
Population growth, 32–33
Primary substrate, 61
Processing, and dumping of waste, 39,
  47–48
Proteins, 15
Proximate analysis, 16
Public concerns, and awareness, 122
Public Health Act, 17, 18
Public health concerns, 121
Public health hazard, 4
Public recognition, 84
Pyrolysis, 80–81

R

Rag pickers, 8, 50–51
RDF, *see* Refuse derived fuel (RDF)
Recyclables, waste, 12
Recyclables bins, 147
Reductionist concept, 126
Red wigglers, *see* Lumbricus rubella
Red worms, *see* Lumbricus rubella
Refuse derived fuel (RDF), 48,, 79–80
Residential waste, *see* Household waste
Residual waste bins, 146
Resource scarcity, 121
Reuse, and recycling, 81–82

S

"The Sanitary Condition of the
      Labouring Population", 17
Sanitary landfilling, 84–88
  biogas recovery from, 87–88
  leachate collection system, 85–87
  methods of, 84–85
    area/ramp method, 85
    trench method, 85
    valley, and ravine area method, 85
Scholl Canyon model, 110–112, 150
Seasonal variation, 35
Secondary mesophilic phase, 91
Secondary storage, 44–46
Semibatch digester, 117
Shell Canyon model, 148
Shred, and burn system, 79–80
Simplified process system, 80
Six M's, 20

Size, of waste, 15
Soft systems thinking, 128
Soil filters, 66
Solid hydrolysis, 103–105
    enzyme-catalyzed kinetics, 104–105
    first-order waste, kinetics, 103–104
Solids retention time (SRT), 101
Solid wastes; *see also* Waste
    collection of, 43–44
    definition, 11
    disposal of, 48–49
    management, 1
    processing of, 47–48
    transportation of, 46–47
SoS, *see* System-of-systems (SoS)
        approach
SRT, *see* Solids retention time (SRT)
Standard-rate single-stage digester, 75
Static, windrows, 67
Stirring, 68
Stoichiometric equation, 112
    for aerobic waste degradation, 91–92
    for anaerobic waste degradation,
        and estimation of theoretical
        methane yield, 96–97
Street sweeping, 31
Street waste pickers, 51
Sulfur compounds, 66
Sweeping of street, and cleaning of
        surface drains, 44
Syntrophy, 95
System-of-systems (SoS) approach,
        129–130
Systems approaches, in municipal solid
        waste management, 119–132
    analysis techniques, 131–132
    centralized and decentralized, 130
    development drivers for, 120–125
        in high-income countries, 120–122
        in low- and medium-income
            countries, 122–125
    disciplinarity and
            multidisciplinarity, 126–127
    engineering principles, 127–129
        approaches, 128–129
        definition, 127–128
        thinking, 128
    integrated, 126
    need for, 125–126

system-of-systems approach, 129–130
Systems assessment tools, 131
Systems engineering models, 131

T

Temperature, 63–64, 74, 98
Terms, of reference, 139
Theoretical methane yield, 96–97
Thermal processing, of municipal solid
        waste, 75–81
    incineration, 76–79
    pyrolysis, 80–81
    refuse-derived fuel, 79–80
Thermophilic organisms, 63
Thermophilic phase, 91
Three-dimensional (3-D) mapping, 147
Top-down design, 129
Transdisciplinary approach, 127
Transfer station, *see* Material processing
        facility
Transition phase, 99
Transportation, MSW, 39, 46–47, 56–58
    optimization of, routes, 57–58
    technical requirements of, vehicles,
        56
    transfer stations, 57
    types of, vehicles, 57
Trench method, 85
Triangular model, 114–115, 150
Tumbling, 68
Turned, windrows, 67

U

UCs, *see* Urban councils (UCs)
UK Environment Agency, 88
ULB, *see* Urban Local Bodies (ULB)
Ultimate analysis, 16
Ultimate disposal methods, 82–88
    carbon storage in landfill, 88
    environmental impacts of landfilling
        and their control, 83–84
    sanitary landfilling, 84–88
        biogas recovery from, 87–88
        leachate collection system, 85–87
        methods of, 84–85
    types of landfill, 82–83
        for milled waste, 83

for mixed waste, 82–83
monofills for specialized waste, 83
Urban councils (UCs), 40
Urbanization, 33, 123
Urban Local Bodies (ULBs), 40, 45, 46
US Environmental Protection Agency
(USEPA), 110–112, 150
USEPA, *see* US Environmental
Protection Agency (USEPA)
USEPA Landfill Gas Emissions Model
(LandGEM), 110–112, 150
USEPA Landfill Monitoring Manual, 87

V

Valley, and ravine area method, 85
Vehicle routing, models for, 147–148
Vermicomposting, 68–71
VFA, *see* Volatile fatty acids (VFA)
VOCs, *see* Volatile organic compounds
(VOCs)
Volatile fatty acids (VFA), 106
Volatile matter, 16
Volatile organic compounds (VOCs), 66
von Bertalanffy, 128

W

Waste degradation, kinetics of, 89–117
aerobic, 89–93
factors affecting, 92–93
initial mesophilic phase, 91
secondary mesophilic phase, 91
stoichiometric equation for, 91–92
thermophilic phase, 91
waste degradation, model, 109–110
anaerobic, 93–99
acetogenesis stage, 94–95
factors affecting, 97–99
hydrolysis stage, 93–94
methanogenesis stage, 95–96
stoichiometric equation for,
and estimation of theoretical
methane yield, 96–97
in anaerobic digesters, 100–101
feedstock/substrate delivery
controlled, 101
kinetics, 115–117

moisture controlled, 101
temperature controlled, 101
landfill gas generation, 110–115
IPCC default method, 113–114
IPCC First-Order Decay model,
113
Scholl Canyon model and USEPA
LandGEM, 110–112
triangular model, 114–115
process, 89
reaction kinetics
anaerobic, kinetic model, 102–109
sequence in landfills, 99–100
Waste
agricultural, 32
characteristics
chemical, 15
physical, 14–15
collection of, 39, 43–44
commercial, 29
construction and demolition, 31
definition, 11
hierarchy, 121
household, 29
inert, 12
institutional, 29
from municipal services, 11
organics, 12
processing and dumping of, 39
recyclables, 12
reducing, 80–81
stream assessment, 12–13
street sweeping, 31
transport and transfer of, 39
Waste-to-energy (WtE)
combustion, 76–77
concept, 21
Wedge system, 69
Wet anaerobic process, 72
White, I. D., 127
Windrows, 67, 69
World Bank, 5, 36
WtE, *see* Waste-to-energy (WtE)

Z

Zero waste and zero landfill, concept
of, 18

9 780367 574284